Silicon, From Sand to Chips 1

Silicon, From Sand to Chips 1

Microelectronic Components

Alain Vignes

WILEY

First published 2024 in Great Britain and the United States by ISTE Ltd and John Wiley & Sons, Inc.

ISTE Ltd
27-37 St George's Road
London SW19 4EU
UK

www.iste.co.uk

John Wiley & Sons, Inc.
111 River Street
Hoboken, NJ 07030
USA

www.wiley.com

Any opinions, findings, and conclusions or recommendations expressed in this material are those of the author(s), contributor(s) or editor(s) and do not necessarily reflect the views of ISTE Group.

Library of Congress Control Number: 2023950040

British Library Cataloguing-in-Publication Data
A CIP record for this book is available from the British Library
ISBN 978-1-78630-921-1

Contents

Preface

At the beginning of the 20st century, silicon "metal" was used as an alloying element for steels with electrical properties. The year 1906 saw the first application of crystalline silicon as a component of electromagnetic wave detection circuits in radio receivers, competing with galena.

Research carried out during the Second World War on silicon and germanium, the materials used in the components (point-contact diodes) of radar receiver circuits for aircraft detection and tracking, revealed that these materials are semiconductors whose basic characteristic is the control of electrical conductivity through doping. This characteristic prompted the search, after the Second World War, for solid components to replace "triodes" (vacuum tubes). This quickly led to the invention of the transistor.

The invention of the transistor is the founding act of the digital revolution (of the information society in which we live).

Germanium then silicon are the first two materials that enabled the invention of the transistor and the initial development of computers, while silicon dethroned germanium to produce the "MOSFET" (metal–oxide–semiconductor field-effect transistor), the basic component of integrated circuits: microprocessors and memories, the building blocks of computers.

But these components require materials (germanium and silicon) of extraordinary purity and perfect crystallinity. The purification of basic materials to purities of up to 11N, the production of single crystals of germanium, then silicon, the manufacture of components (based on transistors) and their miniaturization have posed problems of a complexity rarely encountered in the development of manufactured products.

These are the same properties and characteristics that have made silicon the material of choice for converting solar energy into electricity and for photographic sensors.

Silicon's exceptional mechanical properties, combined with its electrical properties, make it the material of micro-electro-mechanical systems (MEMS), the key components of "intelligent objects".

In 2018, there were no materials on the horizon that were likely to dethrone silicon as the material of choice for microelectronics and optoelectronics alike. According to Gérard Berry: "Silicon is not dead, far from it".

This book is aimed at readers who want to know and understand how it was possible to go from the ENIAC computer, built during the Second World War, to calculate shell trajectories, 30 m long and 2 m high, with 17,468 triodes (vacuum tubes) and capable of executing 5,000 additions and subtractions in 1 s, to centimetric microprocessors with 20 billion transistors, processing power (number of instructions processed per second) of several gigahertz, making up the basic components of the individual computer, which is the size of a thin book.

To this end, this book, by tracing the history of discoveries, inventions, innovations and technological developments in materials, components, integrated circuits and memories, presenting the physical bases of their operation, and focusing on the materials and technologies used to make these components, attempts to answer the following questions:

– What specific properties (characteristics) – electrical, physicochemical, mechanical – are behind the successive dominance of silicon, then germanium, then silicon again in the development of microelectronics, the dominance of silicon in the conversion of solar energy into electricity, the dominance of silicon as the basic material for electromechanical microsystems?

– What properties (purity, crystallinity, doping) had to be imparted to the material, and how were they obtained to achieve the performance achieved by these components today?

– What processes had to be developed to produce these components, and then to meet the demands of miniaturization, enabling the high-speed data processing performance we are seeing today, efficient conversion of solar energy into electricity, etc.?

– Who were the architects of this epic? According to Gérard Berry[1], "its extraordinary success (that of silicon) is clearly to the credit of semiconductor materials physicists, who made technological advances that required enormous imagination and skill to overcome all the obstacles".

Until 1942, silicon extracted from silica (SiO_2) and germanium extracted from sulfide (GeS_2) were considered as metals. The semiconductors known at the time were chemical compounds: oxides (Cu_2O) and sulfides (galena PbS), composed of a metal and a metalloid (oxygen or sulfur), whose basic characteristic was the increasing variation of their conductivity with temperature, whereas the conductivity of metals decreases with increasing temperature. It was not until the summer of 1942 that it was recognized that purified silicon and germanium were not metals, but semiconductors.

This book is divided into two volumes. Volume 1 is devoted to basic components (diodes and transistors).

Chapter 1 presents (1) the work that led to the extraction of silicon from silica and its purification and the discovery, extraction and purification of germanium; (2) the basic physical characteristics of semiconductors made from these two materials, knowledge of which is essential for understanding how components work.

Chapters 2–6 of Volume 1 present the basic components (diodes, transistors) in the chronological order of their discovery/invention, and the technological developments required for their realization.

Each chapter includes a presentation of the component, how it works and its basic functions, followed by the history of the research and development that led to its invention and production. The physical basis of its operation is presented in the appendicies of each chapter. The technologies used to satisfy the requirements of purity and crystalline perfection of the base material are presented chronologically, as are the technologies used to produce the components and the evolutions required by their miniaturization. The industrial development of the first components is presented according to their importance for subsequent developments.

Volume 2 is devoted to "chips, optoelectronic components and MEMS".

Chapters 1 and 2 present microcomputer integrated circuits and memories.

Chapter 3 presents the silicon thin film transistor TFT, which led to the development of flat-panel liquid crystal displays.

---

1 Berry, G. (2017). *L'Hyperpuissance de l'informatique*. Odile Jacob, Paris, p. 88 and 401.

Chapters 4 and 5 present silicon optoelectronic components. These include solar cells for converting solar energy into electricity and photoelectric image sensors for digital cameras, which have revolutionized astronomy and medical imaging.

Chapter 6 presents microelectromechanical systems (MEMS), the exceptional mechanical properties of silicon that have enabled their development, and the specific technologies developed for building structures with moving parts.

Many English and American books present the "history of semiconductors". Compared with the reference works cited in the reference lists, this book presents not only the historical aspects, but also the recent technological developments that have enabled the current performance of microprocessors, memories, solar cells and electromechanical microsystems. The book is based on numerous works by historians and original publications.

The author would particularly like to thank Professors Jean Philibert and André Pineau.

December 2023

References

Burgess, P.D. (n.d.). Transistor history [Online]. Available at: https://sites.google.com/site/transistorhistory.

Computer History Museum (n.d.). The silicon engine timeline [Online]. Available at: www.computerhistory.org.

Hu, C. (2009). *Modern Semiconductor Devices for Integrated Circuits*. Pearson, London.

Krakowiak, S. (2017). Éléments d'histoire de l'informatique. Working document, Université Grenoble Alpes & Aconit, CC-BY-NC-SA 3.0 FR.

Lazard, E. and Mounier-Kuhn, P. (2022). *Histoire illustrée de l'informatique*. EDP Sciences, Les Ulis.

Lilen, H. (2019). *La belle histoire des révolutions numériques*. De Boeck Supérieur, Louvain-la-Neuve.

Lojek, B. (2007). *History of Semiconductor Engineering*. Springer, New York.

Mathieu. H. (2009). *Physique des semi-conducteurs et des composants électroniques*, 6th edition. Dunod, Paris.

Nouet, P. (2015). Introduction to microelectronics technology. Working document, Polytech Montpellier, ERII4 M2 EEA Systèmes Microelectronics.

Orton, J.W. (2004). *The Story of Semiconductors*. Oxford University Press, Oxford.

Orton, J.W. (2009). *Semiconductors and the Information Revolution: Magic Crystals that made IT Happen*. Elsevier, Amsterdam.

Riordan, M. and Hoddeson, L. (1997). *Crystal Fire: The Invention of the Transistor and the Birth of the Information Age*. W.W. Norton & Company, New York.

Seitz, F. and Einspruch, N.G. (1998). *Electronic Genie: The Tangled History of Silicon*. University of Illinois Press, Illinois.

Sze, S.M. (2002). *Semiconductor Devices: Physics and Technology*. Wiley, New York.

Verroust, G. (1997). Histoire, épistémologie de l'informatique et révolution technologiques. Course summary, Université Paris VIII, Paris.

Ward, J. (n.d.). Transistor museum [Online]. Available at: transistormuseum.com.

Introduction

The Digital Revolution

The "digital revolution" is also known as the "computing or IT revolution". These expressions reflect "a radical transformation of the world we are witnessing today".

The first term refers to the binary digitization of texts and numbers, as well as images, sounds and videos, using sequences of symbols. This makes it possible to store images, sounds, etc., and transmit them, replicate them, analyze them and transform them using digital computers (Abiteboul and Dowek 2017, p. 29).

The second expression, "the computing revolution", refers to the science and technique of processing digitized information using algorithms. According to Berry (2017, p. 25), "Computing is the conceptual and technical engine of the digital world. The computer is the physical engine".

The "birth certificate of the digital revolution" is Claude Shannon's 1937 master's thesis, *A symbolic analysis of relay and switching circuits* (1938). This thesis relied on the theory of the Englishman George Boole (*An Investigation of the Laws of Thought*, 1847), which established the link between calculus and logic and where the basic logical functions "AND", "OR" and "NOT" were treated as arithmetic operations, taking the value 0 or 1, depending on whether the proposition was true or false.

The master's thesis of Claude Shannon[1] was the result of an internship at Bell Labs[2], where he observed the power of telephone exchange circuits that used

1 Claude Shannon is also the father of the information theory formulated in 1950, described by the journal *Scientific American* as the "Magna Carta of the information age". (Collins 2002; Berry 2017, p.52).
2 Bell Labs: Bell Telephone Laboratories, a subsidiary of ATT (American Telephone and Telegram Company) with a monopoly on telephone and telegraph transmissions. Its subsidiary, Western Electric, produces components for telephone exchanges.

electromechanical relays (switches)[3] to route calls and imagined that electrical circuits could perform these logical operations using an on-off switch configuration.

The first demonstration of the feasibility of executing logic functions using a device made up of two electromechanical relays was carried out in 1937 by George Stibitz of Bell Labs; this led to the construction in 1939 of the first CNC (complex number calculator) (400 electromechanical relays), capable of opening and closing 20 times a second, executing complex number multiplication and division operations. This was followed by five other models. "Stibitz's calculator demonstrated the potential of a relay circuit to do mathematics in binary, process information, and manipulate logical procedures" (Isaacson 2015, p. 93).

The "digital" revolution is the third major revolution in human history. The first was the agricultural revolution 8,000 years ago. The second was the "industrial revolution" of the 19th century.

The technology at the heart of this third revolution, also known as the "second industrial revolution", is microelectronics[4]. In 1979, the US National Academy of Sciences published a report[5] entitled "Microstructure, Science, Engineering and Technology", which stated: "The modern era of electronics has ushered in a 'second' industrial revolution, the consequences of which may be even more profound than those of the first". According to Ian Ross, President of Bell Labs from 1979 to 1991: "The semiconductor odyssey produced a revolution in our society at least as profound as the total industrial revolution. Today electronics pervades our lives and affects everything" (Ross 1997).

I.1. Microelectronics components

In 1903, Arthur Fleming invented the diode (vacuum tube), a current rectifier, and in 1906, Lee de Forest invented the triode (vacuum tube) by adding a grid between the diode's cathode and anode. As well as rectifying the current, this allowed weak currents induced by electromagnetic waves to be amplified, hence the development of

---

3 The electromechanical relay (a switch that opens and closes by electrical means, such as an electromagnet), with its two states open and closed, was the ideal component for representing the two states of binary numbering (0 and 1) and logic (true or false). With a binary machine, arithmetic operations and logical operations can be processed in the same way.

4 Microelectronics refers to all the technologies used to manufacture components that use electrical currents to transmit, process or store information.

5 Quoted in the brochure "La microélectronique : bilan et perspectives d'une technologie de base", Siemens Aktiengesellschaft, Berlin and Munich, 1984. Translation of the book *Chancen mit Chips*, 1984.

radio receivers: a small variation in the signal on the grid resulted in an amplification of the cathode-anode current. In addition, a sudden variation in the signal applied to the grid switched the triode on or off, enabling it to function as a switch. The triode is also capable of self-oscillation, hence its use in radio transmitters.

The invention of the bipolar transistor in 1948 by William Shockley (Nobel Prize winner), a solid-state device capable of performing the same functions (amplification of weak currents and switching), but much faster, ushered in the era of the digital revolution.

Like transistors, triodes work by controlling a current of electrons, which can either be amplified or interrupted and reignited. These components function like a switch that can be set to 0 or 1 on command, thus performing logic functions. But with triodes, switching times are much longer and the permissible frequencies much lower than in solid-state components, because these variables are linked to the time taken for the electrons to cross the distance between the cathode and the anode (around 1 mm); whereas, in a transistor, the distance traveled by the electrons between the emitter and the collector is less than 1 μm, down to around 20 nm.

Before the invention of the transistor, prototype "computers" had been built with triodes, the ENIAC during the Second World War, then with solid diodes (made of germanium) combined with triodes. Diodes can only be used to create logic circuits (OR and AND gates). They cannot restore the signal at the output of a gate, hence the presence of triodes to restore the signal, enabling cascades of gates to be created, and hence logic circuits.

The discovery of silicon N and silicon P[6], at the beginning of the Second World War, in other words of the effect of doping on the conductivity of silicon and therefore its control, and the discoveries of the rectifier effect and the photoelectric effect presented by the solid-state PN diode[7] by Rüssel Ohl, led to the invention of the bipolar transistor (with PN junctions) in 1949 by William Shockley. The development of circuits made up of solid-state diodes and transistors producing NAND and NOR logic gates and all the universal logic functions by combining one or the other, with the added feature of restoring the signal at the output of each gate, thus enabling cascades of logic gates, led to the development of integrated circuits, invented in 1958–1959 by Jack Kilby (Nobel Prize winner) and Robert Noyce.

The development of the silicon-based field-effect MOSFET transistor, designed by William Shockley in 1945 and by Dawon Kahng and Martin Atalla in 1960, because

---

6 Silicon N (silicon doped with phosphorus) with n-type conductivity; P silicon (doped with boron) with p-type conductivity.

7 PN diode: a component made up of two regions, N and P, joined together along a flat surface.

of its miniaturization capacity, enabled the development of integrated circuits: memories and microprocessors. Microprocessors were the ultimate innovation in the digital revolution, enabling the development of the personal computer. According to Reid (1984), "A new era in electronics had begun".

The miniaturization of components down to the nanometer scale is delivering high performance in terms of information processing speed and substantial savings in power consumption. The number of transistors has risen from 2,400 for the Intel 4004, the first integrated microprocessor, to around 20 billion for today's largest graphics processors (2017). The processing power of a microprocessor (the number of instructions a microprocessor is capable of processing per second) rose from a few megahertz in the early 1980s to several gigahertz in the early 2000s. This clock frequency (as it is known) is directly linked to the switching speed of the microprocessor transistors. We can only marvel that an astronomical set of phenomenally fast electronic components as simple as switches could be the basis of humanity's third revolution.

I.2. Microelectronics materials

These "components" require materials of extraordinary purity and perfect crystallinity to obtain very specific electronic characteristics, as well as completely new technologies for manufacturing transistors and integrated circuits (a list of which is given in the Appendix).

It was the availability and technological mastery of two materials, germanium and silicon, which were virtually unknown at the beginning of the 20th century, with the appropriate electronic characteristics, that enabled the invention of the transistor and conversion of solar energy into electricity.

The purification of base materials to purities of up to 11N (99.999999999), the production of perfectly crystalline single crystals of germanium and then silicon, enabling the conductivity of these materials to be controlled by doping, and the manufacture of components and their miniaturization have posed problems of a complexity rarely encountered in the development of manufactured products (Queisser 1998).

It was with germanium (on purified, coarse-grained (quasi monocrystalline) wafers that were available) that power amplification was first observed in December 1947 on a device made by John Bardeen and Walter Brattain (Nobel Prize winners), which was named the "point contact transistor" (Bardeen and Brattain 1948). This invention led to the development of a process for obtaining single crystals of germanium by Teal (1976). The successful purification and manufacture of germanium single crystals and the development of the bipolar transistor established germanium as the basic material for transistors. In 1952, Ralph Hunter, in a speech as President of the Electrochemical Society of the United States, predicted: "A revolution in the electronics industry as a result of the development of germanium". Germanium transistors were manufactured until 1961. The CDC 1604 and IBM 1401 computers marketed in 1960 were made using germanium transistors.

In 1952, following the successful manufacture of silicon single crystals and of a PN junction in a single crystal, again by Gordon Teal, whose properties were superior to those of the germanium PN junction, "silicon immediately became a rival to germanium" (Leamy and Wernick 1997). Given the difficulties in obtaining "electronic" silicon, silicon very gradually became the preeminent material for transistors, under pressure from the military, who were virtually the only customers at the time – particularly for the temperature resistance of silicon diodes and transistors up to around 150°C.

When the first silicon MOSFET transistor was produced in 1959, silicon's supremacy became total, thanks to the qualities of its oxide and its high thermal dissipation. Since the 1970s, silicon MOSFETs have been the basic components of integrated circuits and computer memories.

In 1951, Heinrich Welker (Nobel Prize) began studies on compounds with the same structure as silicon and germanium, such as gallium arsenide GaAs, revealing their semiconductor characteristics. It was not until 1978 that it was shown that a gallium arsenide component was twice as fast as the same silicon component under the same conditions (Welker 1976). Nevertheless, this factor of 2 did not convince manufacturers to abandon silicon, thanks to its two advantages: its high heat dissipation and its mastered technology (Bols and Rosencher 1988).

I.3. The driving forces behind the development of microelectronics and computer components

Research studies carried out in England and the United States from the start of the Second World War on the reception of radar electromagnetic waves by the "point contact diode" were the first driving force behind the development of silicon and germanium, and marked the first victory of this solid component over vacuum tubes.

The second driving force behind the development of microelectronics was the desire of Bell Labs[8], from the end of the war, to find a solid substitute for the

---

8 "The research group established at Bell Labs in the summer of 1945 had a long-term goal of creating a solid state device that might eventually replace the tube and the relay" (Ross 1997).

triode lamps used as amplifiers along telephone transmission lines and for the electromechanical relays in their ATT telephone exchanges[9].

It was the discoveries of silicon N and silicon P, of the property of rectifying an electric current through a unidirectionally solidified silicon ingot, constituting a PN diode, and of the photovoltaic effect presented by this ingot in 1940, that were at the origin of Bell Labs' adventure in microelectronics. When these remarkable properties of a silicon ingot were brought to the attention of the Bell Labs director, Mervin Kelly considered this discovery of great value to the electronics industry, and decided that absolute secrecy should be preserved until in-depth studies revealed its full power: "It was too important a breakthrough to bruit about".[10] The studies were resumed in 1945.

In the summer of 1945, as reported by Ian Ross, Kelly set up a research group with the following objectives: the fundamental study of semiconductors, concentrating on germanium and silicon, materials which were beginning to be well known, and, in the long term, the creation of a solid-state component constituting an amplifier "to replace triodes (vacuum tubes) and constituting a switch to replace the electromechanical relays of telephone exchanges".

This research, in 1947 and 1949, led to the invention of the point contact transistor and the bipolar transistor with PN junctions.

In 1950, according to Ian Ross, Bell Labs researchers realized that, given the characteristics of transistors, their size and low energy consumption, it was not the replacement of vacuum tubes that should be sought, *but their use as components of logic circuits*[11].

---

9 In the 1950s, a great deal of research was undertaken to apply vacuum tubes to telephone exchange switching. The results were unsuccessful. Until 1974, semi-electronic switches ("space switching") were installed: hybrid switches whose control system is entirely electronic, but which operate on a connection network that is still mechanized, circulating purely analogue conversation currents. The first experimental telephone exchange to use an entirely electronic switching system based on microprocessors, known as "time or digital switching", was set up by the CNET in Perros-Guirec, France, in 1970. These switches constitute the real revolution in modern telecommunications (Caron 1997, p. 295).

10 "The goal for the group, following Mervin Kelly's instructions, was to determine whether it was possible to develop a practical semiconductor triode" (Seitz and Einspruch 1998, p. 164).

11 Ross quoted Bob Wallace: "Gentlemen, you've got it all wrong. The advantage of the transistor is that it is inherently a small size and a low power device. This means that you can pack a large number of them in a small space without excessive heat generation and achieve low propagation delays. And that is what we need for logic applications. The significance of the transistor is not that it can replace the tube but that it can do things that the vacuum could never do". And according to Ross (1997), "And this was a revelation to us all".

As soon as the reproducible manufacture of transistors became possible, in the mid-1950s, "replacing vacuum tubes in as many applications as possible became the objective".

Transistor specimens were entrusted to various Bell Labs engineers with the task of developing applications. John H. Felker was one of them. Felker (1951) showed that the transistor could be used as a component of logic circuits. This potential use of the transistor was presented by Felker to the companies that had acquired the "Western Electric" license. According to McMahon[12], "none of us imagined the revolution that would take place over the next forty years", "even at IBM" according to Rick Dill[13].

Following this presentation, in 1951, the Air Force asked Bell Labs to develop a computer, the TRADIC (transistorized airborne digital computer), which was entrusted to Felker. This resulted in the successive production of four TRADIC computers, of which the Leprechaum version, operational in 1956, was the first fully transistorized computer based on logic circuits made up of bipolar germanium transistors (Irvine 2001).

Most of the discoveries, inventions and technological developments relating to transistors, solar cells and digital photography between 1947 and 1970 were made by Bell Labs researchers (see Table I.1 in the Appendix in this chapter).

Nevertheless, as we shall see, the inventions and technological developments of Bell Labs were not always followed by industrial development and production by Western Electric. There was a good reason for this: ATT, which had a monopoly over telephone and telegraph transmissions in the United States by court order under the anti-trust laws, was only authorized to produce electronic components for its own needs and had to inform the entire electronics industry of any discoveries that might be of interest to it. Therefore, after the first bipolar transistor was produced in 1950, Western Electric began to grant manufacturing licenses to companies producing diodes, triodes (vacuum tubes), etc., "licensing the rights to manufacture transistors for a \$25,000 fee", and for these licensees, Bell Labs organized a Transistor Technology Symposium in April 1951.

---

12 "First, J.H. Felker fascinated us with the applicability of transistors to high-speed digital computers. He stated that the prime objective was practically infinite reliability, closely followed by low power consumption (hence minimal heat removal), small size and minimal weight. However, none of us imagined the semiconductor revolution that was really to take place over the next forty years" (McMahon 1990) (Hughes Aircraft Company).

13 "Everyone on the early transistor business saw analog and communications circuits as the most important thing. In 1954, computers were not important to the electronic world", Rick Dill (IBM) (Dill 1954).

The third driving force was the interest shown by a number of industrial companies who foresaw the importance of these inventions. This was as early as 1948, with the publication of the discovery of the point contact transistor, companies that had been heavily involved in the development and production of germanium diodes during the Second World War: General Electric, Sylvania, RCA, CBS and IBM. Subsequently, other vacuum tube manufacturers who had acquired the Bell license, such as Raytheon, Philco, Telefunken and Siemens, and companies set up by researchers or engineers, who moved from one company to another, produced components and then integrated circuits. In 1958, there were 70 diode and transistor manufacturers in the world, the vast majority of these in the United States (Morton and Pietenpol 1958).

The first commercial computers to use bipolar transistors as logic circuit components appeared in 1956 with the Philco S-2000 and 2600 computers.

Three companies – Texas Instruments (TI), founded in 1952, Fairchild Semiconductor, founded in 1957, and Intel, founded in 1968 – took over from Bell Labs, both in the development and industrial production of the ultimate components.

The inventions and achievements of the integrated circuit in 1958–1959 were due to Jack Kilby of Texas Instruments and Robert Noyce of Fairchild Semiconductor and their collaborators. This invention paved the way for the creation of the "microprocessor" by Intel, a company founded by Fairchild Semiconductor defectors Noyce, Grove and Moore (author of "Moore's Law" in 1969). This was a universal integrated circuit that integrated all the functions of a computer's central processing unit, capable of following programming instructions. In November 1971, Intel presented the Intel 4004 microprocessor.

At the same time, another major driving force behind the development of microelectronics, as with many other major innovations, was the needs of the military or prestige of the state.

The civil space program and the military program to build balistic missiles boosted demand for transistors. The state organizations responsible for these programs financed the companies mentioned above.

The Polaris sea-to-ground ballistic missile program in 1956, then the Minuteman ground-to-ground missile program at the end of the 1950s, for their on-board guidance system, the Vanguard and Explorer earth satellites, launched in 1958. For their transmissions, the Apollo program, at the beginning of 1960, endowed with 25 billion dollars, gave a real boost to research into integrated circuits and computers[14].

I.4. The material of solar energy

Diodes made of silicon, germanium and other semiconductors convert photons into electrons.

The photovoltaic effect, presented by a silicon ingot forming a PN diode, was discovered by Russell Ohl of Bell Labs in 1940 (Ohl 1946).

The solar cell development program began in 1952. On April 26, 1954, Bell Labs announced the manufacture of silicon solar cells using the diffusion doping process. It was the development of this doping process for solar cells that ensured the development of transistors (Chapin et al. 1954).

The space program was the driving force behind the development of solar cells; the first use of solar cells was on the Vanguard 1 satellite, launched on March 17, 1958, to power a radio transmitter. The system operated for 8 years. The space program stimulated (financed) a great deal of research and a veritable cell production industry.

The energy crisis of 1974–1975 sparked renewed interest in solar cells. Silicon is the material of a major energy source (solar energy): 99.4% of solar panels are based on silicon, and 0.4% on CdTe and GaAs.

Solar Impulse 2, the fragile aircraft with its huge wings covered with solar panels (11,628 ultra-fine monocrystalline silicon photovoltaic cells (each 135 μm thick)), is the symbol of the progress made in just a few years in the field of materials and renewable energies.

I.5. The material of digital image sensors

Silicon photoelectric image sensors, whose invention by W.S. Boyle and G.E. Smith, also of Bell Labs, in 1970 won them a Nobel Prize, have made digital

---

14 "The decision of President Kennedy in 1961 to mount an intensive space programme, with the in-tention to put a man on the moon in 1970 kick-started a technological revolution, certainly no other country ever received a comparable boost. Given the modest lifting capability of current US rockets weight was a vital factor and all electronics must therefore be transistorized. Ruggedness and relia-bility too were better served by solid-state devices than by the older fragile vacuum tubes. It became clear that the required rocket guidance would demand highly sophisticated computer technology and that such advanced circuitry could only be realised in integrated form" (Orton 2004, p. 99).

photography possible and revolutionized astronomy. They are crucial components of fax machines, cameras, scanners and medical imaging (Boyle and Smith 1970).

I.6. The material of micro-electro-mechanical components (MEMS)

A MEMS "sensor" or "actuator" is an essential part of what we call intelligent objects, since it is thanks to them that we can obtain information linked to our environment and vice versa. A series of technological breakthroughs and industrial bets have helped to explode a market that continues to evolve (Vigna 2013).

The development of these microsystems has been made possible by the availability of a material, silicon, with its exceptional electrical and mechanical properties, and by the development of specific miniaturization technologies for this material.

I.7. The role of "metallurgists"

While the invention of the transistor can undoubtedly be attributed to three physicists: John Bardeen, Walter H. Brattain, William B. Shockley of Bell Labs, according to Jack Scaff, director of the transistor laboratory materials from Bell Labs: "The role of metallurgists in these developments was essential" (Scaff 1970).

The purification of basic materials, down to purities of 11N, and the miniaturization of components (transistors) have posed material problems of a complexity rarely encountered in the development of manufactured products.

The appended table lists the main inventions, discoveries and achievements by "metallurgists/chemists"; for instance, the discovery of silicon N and silicon P, the discoveries of the rectifier effect and of the phoelectric effect, presented by a unidirectionally solidified silicon rod by Rüssel Ohl from ATT's Bell Labs in 1940.

It is to Gordon Teal of Bell Labs that we owe, from the discovery of the point contact transistor in December 1947, the recommendation to purify the material and to use single crystals as the base material of transistors, the development of the CZ pulling method to obtain single crystals of germanium initially, then of silicon, the development of processes to obtain the greatest purity of the base material, the development of the doping process for germanium and then silicon, which led to the production of the first bipolar transistor in germanium.

The work of Jack Scaff and Henry Theuerer, and then Calvin Fuller on the diffusion doping process is noteworthy; this became the basic manufacturing process for transistors (later replaced by ion implantation, based on the same principle) (Chapter 5).

The discovery of the masking process by oxidation of silicon, by Derrick and Frosch, enabled the invention, by Jean Hoerni of the Fairchild company, of the bipolar transistor with a planar configuration in silicon. According to Bo Lojek, Hoerni carried out his experiments by working alone, practically at night, without any research budget, taking care not to inform Gordon Moore (Lojek 2007, p. 123).

Atalla's discovered silicon passivation by oxidation, which led to the development of the first silicon MOSFET transistor by Khang and Atalla (Chapter 6).

I.8. The technological keys to the digital revolution

The technological keys to the digital revolution are as follows:

– the discovery of silicon N and silicon P in 1940; in other words, the discovery that the conductivity of silicon, depending on doping with elements such as boron or phosphorus, could be of the n type (by electrons carrying a negative charge) or the p type (by holes, carrying a positive charge), a property also observed for germanium. Further research in 1942 established that these elements were "semi-conductors";

– the discovery of the silicon-based PN diode in 1940;

– the invention of the bipolar transistor in 1949 by W. Shockley solid electronic component, designed to replace the triode as an amplifying element, but also proved to be a switching element: a "real electronic valve";

– the invention of the integrated circuit in 1959: a major innovation consisting of the "monolithic integration", in a single crystal of silicon, of logic circuits made up of several transistors, up to billions, which ushered in an industrial revolution (the second or third);

– purification to unprecedented levels for a material (up to 11N) and the production of perfectly crystalline single crystals for germanium and then for silicon, enabling the conductivity of these semiconductor materials to be controlled by doping;

– the qualities of silicon oxide: stability, insulator (dielectric), diffusion barrier, selective etching by HF (Hydrofluoric acid) (without attacking the underlying silicon), which enabled the creation of the MOSFET field effect transistor and the development of integrated circuits;

– the thermal dissipation of silicon (the thermal conductivity of silicon is twice that of germanium and three times that of gallium arsenide), enabling advanced integration.

I.9. Appendix

Innovations	Companies	Dates	Authors
Controlling Si conductivity by doping	BTL	1942	Russell Ohl and Jack Scaff
Rectifier effect and photovoltaic effect of the silicon PN diode	BTL	1940	Russell Ohl
Point contact transistor	BTL	1947	John Bardeen and Walter Brattain (Nobel)
Bipolar transistor (with PN junctions)	BTL	1948	William Shockley (Nobel)
Germanium (Ge) single crystal	BTL	1950	Gordon Teal
***Grown junction* Ge bipolar transistor**	BTL	1950	Morgon Sparks and Gordon Teal
Ge purification by zone melting	BTL	1951	W.C. Pfann
Ge bipolar transistor made by alloying (*alloy junction transistor*)	General Electric	1951	R.N. Hall and W.C. Dunlap
Silicon bipolar transistor	BTL-TI	1954	Morris Tannenbaum and Gordon Teal
Production of a Si solar cell by boron diffusion	BTL	1954	Calvin S. Fuller
Oxide masking process of Si (*oxide masking*)	BTL	1955	Lincoln Derrick and Carl Frosch
Si planar bipolar transistor	Fairchild	1959	Jean Hoerni
***Oxide surface* passivation of silicon**	BTL	1959	Martin Atalla, Morris Tannenbaum and E. Scheibner
Integrated circuits	TI and Fairchild	1959	Jack Kilby (Nobel) and Robert Noyce
Si MOSFET transistor	BTL	1960	Dawon Khang and Mohamed M. Atalla
Complementary MOSFET transistor (CMOS)	Fairchild	1963	Frank Wanlass, Chih Tang Sah, Moore
***Floating gate MOSFET* (ROM memory)**	BTL	1967	Dawon Kahng and Simon Min Sze
DRAM memory (*one transistor DRAM cell*)	IBM	1968	Robert Dennard
CCD digital image sensors	BTL	1970	Willard S. Boyle and George E. Smith (Nobel)

(BTL: Bell Telephone Laboratories, "Bell Labs"; TI: Texas Instruments)

Table I.1. *Main discoveries, inventions and innovations (adapted from P. Seidenberg,"From germanium to silicon", ethw.org/archives)*

I.10. References

Abiteboul, S. and Dowek, G. (2017). *Le temps des algorithmes*. Le Pommier, Paris.

Bardeen, J. and Brattain, W.H. (1948). The transistor, a semi-conductor triode. *Physical Review*, 74(7), 230–231.

Berry, G. (2017). *L'Hyperpuissance de l'informatique*. Odile Jacob, Paris.

Bols, D. and Rosencher, E. (1988). Les frontières physiques de la microélectronique. *La recherche*, 203, 1176–1186.

Boyle, W.S. and Smith, G.E. (1970). Chargecoupled semiconductor device. *Bell System Technical Journal, B.S.T.J.Briefs*, 49(4), 587–593.

Caron, F. (1997). *Les deux révolutions industrielles du XXème siècle*. Albin Michel, Paris.

Chapin, D.M., Fuller, C.S., Pearson, G.L. (1954). A new silicon p-n junction photocell for converting solar radiation into electric power. *Journal of Applied Physics*, 25(5), 676–677.

Cohen, D. (2015). *Le monde est clos et le désir infini*. Albin Michel, Paris.

Collins, G.P. (2002). Claude E. Shannon: Founder of the information theory. *Scientific American*.

Dill, R. (1954). Germanium alloy transistors [Online]. Available at: http://ibm-1401.info/germaniumalloy.html.

Felker, J.H. (1951). The transistor as a digital computer component. In *Proceedings AIEE-IRE Computer Conference*, Philadelphia, American Institute of Electronic Engineers (February 1952), 105–109.

Irvine, M.M. (2001). Early digital computers at Bell Telephone Laboratories. *IEEE Annals of the History of Computing*, 23(3), 21–42.

Isaacson, W. (2015). *Les innovateurs*. JC Lattès, Paris.

Leamy, H.J. and Wernick, J.H. (1997). Semiconductor silicon: The extraordinary made ordinary. *MRS Bulletin*, 47–55.

Lojek, B. (2007). *History of Semiconductor Engineering*. Springer, Berlin.

McMahon, M.E. (1990). The great transistor symposium of 1951, reprinted from SMEC. *Vintage Electrics*, 2. Prologue.

Morton, J.A. and Pietenpol, W.J. (1958). The technological impact of transistors. *Proceedings of the IRE*, 6, 955–959.

Ohl, R.S. (1946). Light sensitive electric device. Patent, US2402662.

Orton, J.W. (2004). *The Story of Semiconductors*. Oxford University Press, Oxford.

Queisser, H.J. (1998). Materials research in early Silicon Valley and earlier. Semiconductor silicon. *The Electrochemical Society Proceedings Series*, PV 98-1, 4.

Reid, T.R. (1984). *The Chip*. Simon & Schuster, New York.

Ross, I.M. (1997). The foundation of the silicon age. *Bell Labs Technical Journal*, 2(4), 3.

Scaff, J.H. (1970). The role of metallurgy in the technology of electronics materials. *Metallurgical Transactions*, 1(3), 561–573.

Seitz, F. and Einspruch, N.G. (1998). *The Tangled History of Silicon*. University of Illinois Press, Illinois.

Shannon, C. (1938). A symbolic analysis of relay and switching circuits. *Transactions of the American Institute of Electrical Engineers*, 57(12), 713–723.

Teal, G.K. (1976). Single crystals of germanium and silicon – Basic to the transistor and integrated circuit. *IEEE Transactions on Electron Devices*, 23(7), 621–639.

Vigna, B. (2013). La revolution des MEMS, Interview (ST microelectronics). *Paris Tech Review*, 9 December [Online]. Available at: www.paristechreview.com.

Welker, H.J. (1976). Discovery and development of III-V compounds. *IEEE Transactions on Electron Devices*, 23(7), 664–674.

Silicon and Germanium: From Ore to Element

Silicon, a conjecture by Lavoisier in 1789, was isolated by Gay-Lussac in 1808. The extraction of silicon by carbothermic reduction of silica in an electric arc furnace was achieved by Henry Moissan in 1897.

Germanium was discovered and isolated by Clemens Winkler in 1886. Chemical extraction and purification processes from germanium sulfide and purification were developed before the Second World War.

This chapter presents:

– the work carried out to isolate and extract the elements silicon and germanium;

– the industrial production process for "silicon metal";

– extraction and purification of germanium;

– the basic electrical characteristics of semiconductors (silicon and germanium), which are essential for understanding the operation of the components presented in the following chapters.

1.1 Extraction and purification of silicon/discovery and extraction of germanium

1.1.1. *Silicon: from quartz to silicon metallurgy*

Silicon is the most abundant element in the Earth's crust after oxygen, 25.7% of its mass, but in the form of silica, SiO_2.

Silica was known in ancient times as "quartz". Silica was considered an element by alchemists, and then by chemists up to and including Lavoisier. Nevertheless, Lavoisier, the father of modern chemistry, was the first, in 1789, to hypothesize (predict) that silica was the oxide of a "fundamental chemical element".

In his *Traité élémentaire de chimie présenté dans un ordre nouveau et d'après les découvertes modernes*, Lavoisier (1789, p. 192) clarified the concept of an element as a simple substance that cannot be decomposed by any conventional method of chemical analysis, and established the first classification of 55 "simple substances", including oxygen, nitrogen and hydrogen, sulfur, phosphorus, carbon and 17 metals and five "earths", which could not be decomposed. He drew up a "table of simple substances, or at least those that the current state of our knowledge obliges us to consider as such", broken down into a sub-table of simple oxidable and acidifiable metallic substances (ranging from antimony to zinc, including molybdenum and tungsten) and a subtable of simple metallic substances (ranging from antimony to zinc, including molybdenum and tungsten) and a sub-table of simple earthy salineable substances (earths that can react with acids to produce salts), which includes lime, magnesia, barite, alumina (clay, alum earth, alum base) and silica (siliceous earth, vitrifiable earth). "The composition of these earths is absolutely unknown", and on pages 194–195 Lavoisier writes:

> It is to be presumed that earths will soon cease to be counted among simple substances. They are the only substances in this class that have no tendency to unite with oxygen. In this view, earths would be metal oxides oxygenated to a certain extent. Strictly speaking, it would be possible for all the substances to which we give the name of earths to be nothing more than metal oxides that cannot be reduced by the means we use. This is merely a conjecture that I am presenting here.

1.1.1.1. *Electrochemical extraction of metals from their oxides*

In 1799, Volta designed an electric battery, consisting of a stack of alternating zinc and silver discs, each pair separated by a wet felt disc impregnated with a salt solution, providing a continuous electric current. As soon as the Volta was known, many experimenters tried to observe the effect of electric current on chemical compounds, in particular: decomposition by electrolysis.

The pioneers in this field were Wilhelm Hisinger and Jöns Jacob Berzelius, on the one hand, and Humphry Davy, on the other hand. Gay-Lussac and Thénard (1811) summarized their work:

> There are times when the sciences are cultivated more ardently and successfully than others. It is when, after many years without any

remarkable discovery, a dazzling one is suddenly made. What movement was produced by the discovery of the voltaic battery? It appeared, and it was felt that it should make the already famous name of Volta immortal. Batteries were built everywhere. Scientists studied their marvelous effects over and over again... But what was easy to foresee happened; this movement gradually slowed down: it took a new and brilliant discovery to revive it; and it was Davy who had the glory of making it. Hisinger and Berzelius had just proved that the voltaic current decomposed oxides and acids, transporting the oxygen to the positive pole and the radical to the negative pole, and that it decomposed salts by transferring the entire acid to the first of these poles and the base to the other. Fascinated by this law of decomposition, Davy was quick to notice it and apply it in numerous ways. After subjecting acids, salts and many other bodies to the action of the battery, he was led to subject alkalis to it: new and very remarkable phenomena appeared to him. At the negative pole, shiny, metallic substances gathered [...] These were potassium and sodium...

Humphry Davy carried out a systematic study on the decomposition by igneous electrolysis of alkaline-earth bases (lime, baryte) and isolated the corresponding "metals" at the negative electrode to which he gave the names barium, strontium, calcium and magnium (and not magnesium, attributed to metallic manganese), identified as metallic substances ("it is a perfect conductor of electricity"). Tests were carried out on alumina, *flint (silica)*, zirconia and glucine.

From the general tenor of these results, there seems very great reason to conclude that alumine, zircone, glucine and silex are, like the alkaline earths, *metallic oxides*, for on no other supposition is it easy to explain the phenomena that have been detailed. [...] Had I been so fortunate as to have obtained more certain evidences on this subject, and to have procured the metallic substances I was in search of, I should have proposed for them the names of *silicium*, alumium (aluminium), zirconium and glucium (Davy 1808, pp. 352–353).

1.1.1.2. *Extraction of silicon from silica by chemical means*

Silicon was "extracted" from silica by chemical rather than electrochemical means, firstly by Joseph Louis Gay-Lussac in 1808[1], then by Jöns Jacob Berzelius in 1823, and by Henry Sainte-Claire Deville in 1854. But it was the work of Humphry Davy which was responsible for the chemical "isolation" of silicon. The very strong

1 "It was of the utmost importance to obtain large quantities, either to study their properties or to use them in the future as reagents" (Gay-Lussac and Thénard 1811).

attraction of potassium to oxygen observed by Davy led him to try to reduce oxides of the bases lime and magnesia directly by bringing them into contact with potassium.

It was by systematically studying the action of potassium and sodium on metal oxides ("this metal being capable of decomposing all metal oxides") that Gay-Lussac and Thénard isolated silicon from "silicon fluoride gas". By reacting potassium with silicified fluoric acid (fluosilicic acid) at 200–250°C (reaction [1.2]), obtained by reaction [1.1], they obtained "a chocolate-colored, bright-light solid", without purifying it, characterizing it or identifying it as a new chemical element. It was indeed silicon, the reactions taking place being:

$$6FH(g) + SiO_2 \rightarrow H_2 SiF_6 (g) + 2H_2O \hspace{3cm} [1.1]$$

$$K(s) + H_2 SiF_6 (g) \rightarrow K_2 SiF_6 (s) + H_2 \hspace{3cm} [1.2]$$

$$K_2 SiF_6 (s) + 4K (s) \rightarrow Si(s) + 6KF(s) \hspace{3cm} [1.3]$$

This treatise does not contain any chemical formulae, which were moreover unknown, as Gay-Lussac's conclusion shows (Gay-Lussac and Thénard 1811).

"Amorphous" silicon was isolated and characterized in 1823 by Jöns Jacob Berzelius using the same preparation method as Gay-Lussac but eliminating the excess potassium fluate by rinsing in water. Thrown on a filter, well washed and dried, it was silicon in its purest state. The silicon is a dark hazelnut brown, without the slightest metallic sheen. It does not conduct electricity. The characterization was carried out:

by combustion of this residue, silica was formed. The combustion residue, treated with fluoric acid, gave silicon fluoride gas. [...] Since it has neither luster nor the property of conducting electricity in the state in which it was obtained, it is obvious that it cannot be classified as a metal and that its properties are similar to those of boron and carbon.

Berzelius similarly isolated zirconium in 1824 and thorium in 1828 (Berzelius 1824).

Henri Sainte-Claire Deville, in 1854, carried out igneous electrolysis of a soft salt of aluminum and sodium (cryolite: $AlF_6 Na_3$) (containing silicon because it comes from an aluminum ore containing it). He obtained an aluminum "melt" (deposited on the negative platinum electrode) containing silicon, and isolated the silicon by dissolving the aluminum using concentrated boiling hydrochloric acid.

This silicon is in shiny metallic strips and in this state it differs essentially from the silicon of Berzelius. [...] It resists the action of fluoric acid and is dissolved only in a kind of aqua regia formed from fluoric acid and nitric acid [...] However, I do not believe that silicon is a true 'metal': on the contrary, I think that this new form of silicon is to ordinary silicon (obtained by Berzelius) what graphite is to coal. [...] This graphitoid silicon conducts electricity like graphite (Deville 1854).

In 1855, Henri Sainte-Claire Deville obtained crystalline silicon "which is the analogue of diamond" by reducing silicon chloride with sodium. "Silicon therefore differs from metals in every respect" (Deville 1855a, 1855b).

Friedrich Wöhler, in 1856, obtained crystalline silicon by reacting double fluoride of silicon and potassium ($K_2 SiF_6$) with sodium or aluminum (reaction [1.3]) (Wöhler 1857).

1.1.1.3. *Extraction of silicon by carbothermic reduction of silica: "metallurgical" silicon*

The reduction of oxides by carbon had been known for a long time. In 1888, Warren (1888) showed that silicon could be produced "economically" by heating a mixture of silica, carbon and iron.

The discovery of the electric arc is attributed to Sir Humphry Davy in 1801. In 1808, Davy used an electric arc to melt a small quantity of iron. Siemens and Huttington were responsible for the first use of a practical electric furnace. The coal crucible formed one of the electrodes and the current passed through the mass to be melted (Davy 1808). In 1878, William Siemens patented a type of arc furnace.

In his laboratory arc furnace, built in 1892, Henri Moissan reached temperatures of up to 3,000°C. The charge placed in a carbon crucible was heated by the electric arc produced between two carbon electrodes.

The first "preparation" of silicon and a silicon-rich alloy in an electric arc furnace was carried out by Henri Moissan in 1897 (Moissan 1897).

Until Henri Moissan, some oxides were considered to be irreducible by carbon. Moissan, in his electric furnace, systematically studied the reduction of all of the oxides previously considered irreducible by carbon: chromium, manganese, molybdenum, tungsten vanadium, zirconium titanium aluminum. "With silica, the formation of silica vapor (SiO gas) is very abundant, but if the silica is not completely volatilized, the pellet removed from the crucible contains silicon crystals mixed with silicon carbide crystals". It should be noted that for Moissan, as for Henri Sainte-Claire Deville, silicon was a "metalloid" as are boron and carbon.

Industrial production began in 1898 at the Compagnie Générale d'Électrochimie de Bozel, in Savoie, France.

The production of silicon in an electric furnace, using the purest silica available from quartz sand by coke reduction, starting in 1908 by the Carborundum company of Niagara Falls, is described in numerous patents (Tone 1913).

1.1.1.4. *Purification of silicon produced in the arc furnace*

1.1.1.4.1. Purification by contact with a silicate slag

This process, developed by the Carborundum company (Brockbank 1916), produces silicon of 98% purity, before contact with a slag: Si 92.05%, Fe 4.85%, Al 1.80%; after contact with the slag: Si 97.47%, Fe 1.20%, Al 0.33%.

1.1.1.4.2. Purification by leaching

This purification method was developed in 1919 and 1921. It involves "leaching" silicon powders with an acid, which dissolves the metallic impurities segregated at the grain boundaries. Metallic elements have a very low solubility in silicon in the solid state and therefore segregate strongly at grain boundaries during solidification. By grinding the raw silicon mass very finely, the powder obtained is made up of grains (crystals) whose surface is enriched in metallic elements. The elements segregated at the surface of the acid-soluble grains are thus dissolved.

From metallurgical silicon Si 98%, Fe 0.59%, Al 0.45%, Ca 0.80%, the following two leaching purification processes produced purities of 99.85% and 99.95% (Scaff 1946):

– Electrometallurgical Company process (Becket 1921): Si 99.85%, C 0.019%, O 0.061%, Fe 0.031%, Al 0.020% (Ca, Mn, Mg), P 0.011%, B 0.005%.

– Tücker process: Si 99.95%, Fe 0.02%, Al 0.02%, Ca 0.009% (Tücker 1927).

There are three levels of silicon purity depending on its use:

– metallurgical silicon   produced in an arc furnace (99.99% purity) by carbo-thermal reduction of quartz and purification by leaching[2];

– solar silicon (purity 8N/99.999999%), known as SOG-silicon or s-Si (solar grade);

– electronic silicon (purity 11N/99.999999999 %; impurities < 1 ppb), known as EG-silicon or e-Si (electronic grade).

---

2 A presentation of the modern carbothermal silica reduction process can be found in Vignes (2011).

Silicon production processes: metallurgical silicon, solar silicon and electronic silicon are presented chronologically in Chapters 2 (section 2.3.1.2) and 4 (section 4.2.2) of Volume 1 and in Chapter 4 (section 4.2) of Volume 2.

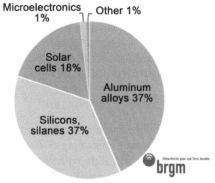

Worldwide consumption: 3.0 Mt

Figure 1.1. *Uses of metallurgical silicon (BRGM 2019). For a color version of this figure, see www.iste.co.uk/vignes/silicon1.zip*

1.1.2. *Germanium*

Germanium is a rare element, with a content of 1.5 ppm in the Earth's crust. Its main application today is in optical fibers and infrared lenses.

1.1.2.1. *Discovery and identification of germanium*

In 1871, Mendeleev published a table predicting the properties of three missing elements, which he called ekabore, ekaaluminium and ekasilicium. The subsequent discoveries of gallium (1875), scandium (1879) and germanium (1886), which could be identified with the three previous elements, would provide a striking vindication of his prediction, right down to the details of the physicochemical properties of these elements.

In 1886, Clemens Alexander Winkler isolated an element, which he called germanium, contained in a newly discovered ore in Germany containing a silver-rich mineral, argyrodite ($4Ag_2S.GeS_2$), composed of silver sulfide (93–94%) and some impurities. By fractional precipitation of solutions of mixed sodium sulfate and impurities, with the addition of hydrochloric acid, Winkler successively isolated antimony sulfide (Sb_2S_3), arsenic sulfide ($As_2 S_3$) and germanium sulfide (GeS_2) (Winkler 1886).

As the elements As, Sb and Bi are in the same column of the classification, Winkler proposed placing germanium between antimony (Sb) and bismuth (Bi) because of their chemical similarity. When Mendeleev became aware of Winkler's discovery, he contacted him. Mendeleev doubted Winkler's hypothesis and suggested that this new element might be an ekacadmium. For Julius Lothar Meyer, on the other hand, who had drawn up a table similar to Mendeleev's in 1870, this new element was ekasilicium. Having great difficulty in trying to determine the atomic mass of germanium, Winkler sent samples to Paul-Émile Lecoq de Boisbaudran, who had discovered gallium, for analysis. Using spectrography, de Boisbaudran obtained an atomic mass value of 72–73, in agreement with Mendeleev's prediction of the position of ekasilicon and its atomic mass of 72 (Burdette 2018).

1.1.2.2. *Extraction and purification of germanium*

In the 1920s and 1930s, research continued on the extraction of germanium and gallium from a sulfide ore (germanite) containing 5–6% germanium, discovered in South Africa in 1922. Germanite is a sulfide with the formula $Cu_{26}Fe_4Ge_4S_{32}$. The process developed by Patnode and Work involved chlorination of the ore at 350°C, fractional distillation to extract the chloride $GeCl_4$, hydrolysis and precipitation of the oxide GeO_2, further conversion to chloride for final purification by distillation (Patnode and Work 1931). In the United States, at Brown University, Professor Charles A. Kraus in 1935 developed a method for the direct extraction of germanium by distillation of germanium sulfide (Johnson et al. 1935). Gordon Teal from Bell Labs, whose key role in obtaining Ge and Si single crystals we shall see later (Chapter 4), wrote his Master's and PhD theses (1931) on germanium compounds (Teal and Kraus 1950).

1.2. Silicon and germanium semiconductors: electrical characteristics

1.2.1. *History: the resistivity of the "metals" silicon and germanium*

1.2.1.1. *Silicon "metal"*

Following the invention of the Volta battery in 1799, many experimenters studied the effect of electric current on metals and chemical compounds.

In 1821, Davy (1821) studied the conductivity of many metals and concluded that their conductivity decreased with increasing temperature (Seitz and Einspruch 1998, p. 50).

Michael Faraday in 1833[3] found that the conductivity of silver sulfide (Ag_2S) increased with temperature (Faraday 1833).

---

3 "I have met with an extraordinary case, which is in direct contrast with the influence of heat upon metallic bodies, as observed and described by Sir Humphry Davy, the substance presenting this effect is the sulphuret of silver. [...] The conducting power increased as the whole became warm and the sulfuret was found conducting in the manner of a metal" (Faraday 1833).

Johann Wilhelm Hittorf, in 1851, confirmed the increasing variation in conductivity with the temperature of the sulfides Ag_2S and Cu_2S (Hittorf 1851).

Before the Second World War, the term semiconductor was used exclusively for compounds such as copper oxide, silver sulfide and lead sulfide (galena), their main characteristic being the increasing variation of their electrical conductivity with temperature (Busch 1989)[4].

It was Alan Wilson, in 1931, who recognized the existence of semiconductors as a specific class of materials. He is the father of the theory of the energy bands of solids (see Figure 1.5). He was the first to explain the difference between metals and insulators, an explanation based on the concept of filled and empty "energy bands" (Wilson 1931).

The first measurements of electrical resistivity as a function of temperature of a sample of metallurgical silicon were carried out by Koenigsberger and Schilling in 1908 (Figure 1.2).

The study looked at three elements with four external electrons and classified as metallic. These curves show that the resistivity of silicon decreases (conductivity increases) with increasing temperature, unlike that of metals.

Nevertheless, in 1931, silicon was still considered a "metal", particularly by Alan Wilson[5], although he noted that, according to measurements (between 4 and 273 K) made by Meissner and Voigt on metallurgical silicon (0.5% Fe, Al, Ca) in 1930, resistivity decreased (conductivity increased) by a factor of 200 when the temperature increased from 4 to 273 K, in agreement with Koenigsberger's measurements and in the opposite direction to that of metals. Wilson (1931) attributed this variation to the "undoubted" presence of oxygen (without further explanation). Oxygen, which is very soluble in silicon, as in titanium and zirconium, after heat treatment, forming SiO_4 complexes, is an electron "donor". The "tested"

---

4 Two articles review the history of semiconductor research (Pearson and Brattain 1955; Busch 1989).

5 "There are substances such as silicon which show a negative temperature coefficient in the impure state, but which are good metallic conductors in the pure state and are therefore to be classified as metals. There are some substances such as germanium which probably belongs to both classes. Recent measurements by Meissner did not confirm the metallic character of silicon though this is almost certainly due to the presence of oxygen in his single crystals" (Wilson 1931).

silicon would therefore be a semiconductor of conductivity n (defined below), the variation with temperature of its resistivity at very low temperatures being consistent with that of an n-doped semiconductor, due to the increase in the concentration of conduction electrons in the freeze-out region (section 1.2.6 and Figure 1.9).

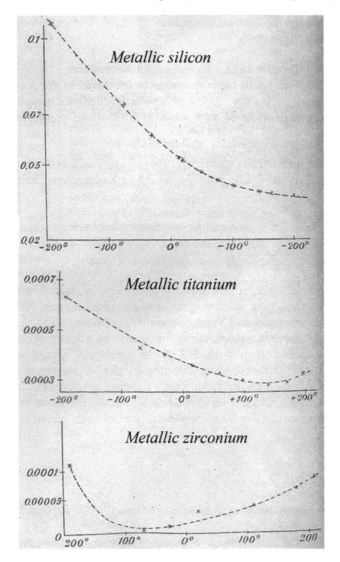

Figure 1.2. *Variations of the resistivity (ohm.cm) of silicon, titanium and zirconium with temperature (Koenigsberger and Schilling 1908)*

1.2.1.2. *Germanium metal*

In 1915, measurements of the conductivity of germanium were carried out by Carl Benedicks (1915) the conductivity of germanium was intermediate between that of silicon and tin.

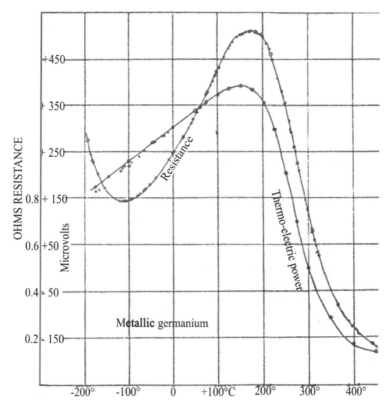

Figure 1.3. *Variation of the resistance (ohms) and thermoelectric power (microvolts) of germanium with temperature (Bidwell 1922)*

Measurements of the electrical resistance and thermoelectric power between –200 and +600°C of a sample of "purified" germanium, described as a "metal", were carried out by Bidwell in 1922 (Figure 1.3), showing different behavior in three temperature ranges. Resistance decreased from –200 to –100°C ("freezing" range), then increased from –100 to +100°C (extrinsic range), then decreased again above 200°C (intrinsic range) (see Figure 1.9) (Bidwell 1922). As we shall see later, all of these results, for both silicon and germanium, were in fact characteristic of an extrinsic semiconductor at low and medium temperatures and of an intrinsic semiconductor at high temperatures (resistivity drops with temperature) (Seitz and Einspruch 1998, p. 127).

It was the basic property of these semiconductors, discovered in 1940, namely, the ability to modulate and control electrical conductivity by specific addition (dopants), over more than 10 orders of magnitude, but within a limited temperature range, that enabled the development of solid components based first on germanium, then on silicon, as a replacement for vacuum tubes.

In the summer of 1942, following research carried out at the start of the Second World War, Professor Lark-Horovitz established that silicon and germanium were not metals, but semiconductors.

Basic characteristics of silicon and germanium as semiconductors[6]:

– the width of the energy gap separating the valence band from the conduction band;

– the generation/recombination of "electron-hole" pairs;

– electrical conductivity, a function of doping and temperature.

1.2.2. *Energy bands*

Silicon (like germanium) is an element in the fourth column of Mendeleev's classification whose four outer electrons form covalent bonds between the atoms of the crystalline structure (Figure 1.4). These electrons are more or less strongly bonded to the atoms in the lattice depending on the width of the respective energy band gap.

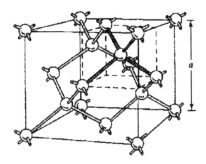

Figure 1.4. *Crystal structure of silicon, a = 0.543 nm*

---

6 A full account of the basic semiconductor characteristics and concepts presented in this chapter can be found in Sze (2002, Chapter 2, p. 17).

Band theory states that the energy (internal energy) (in electronvolt: eV) of an electron in a solid can only take on values within certain ranges called "bands", which are separated by a band of forbidden energy (band gap).

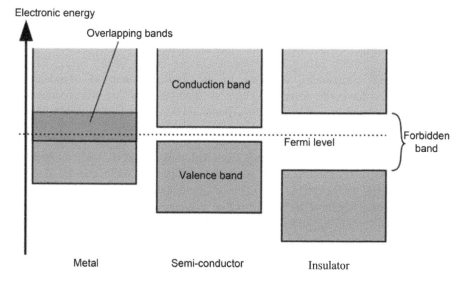

Figure 1.5. *Diagram of the energy bands of metals, insulators and semiconductors: valence band, conduction band and Fermi level. For a color version of this figure, see www.iste.co.uk/vignes/silicon1.zip*

For conductors (metals), the conduction energy band and valence energy bands overlap. Electrons whose energy is higher than the Fermi level circulate freely. These are called conduction electrons. The Fermi level (energy) E_F defines the "work function": the energy that must be supplied to a conduction electron in order to bring it into vacuum (work function) (see Figure 1.7). The Fermi energy is defined in section 1.2.4.

For a semiconductor, as for an insulator, these two bands are separated by an "energy gap E_G". This is the energy (thermal energy, for example) that must be supplied to a "valence band electron" for its energy to reach the lower limit of the conduction band. We say "to make the electron pass from the valence band to the conduction band". The difference between a semiconductor and an insulator is the width of this energy band gap, which gives each its respective properties.

For an insulator, this value is very high (around 6 eV for diamond, for example), "the electrons cannot pass from the valence band to the conductivity band": the electrons do not circulate in the solid.

For semiconductors, an energy input equal to the width of the energy band gap E_G is sufficient "to make an electron move from the valence band to the conduction band" and become a conduction electron (free to move).

Silicon and germanium, elements in the fourth column of Mendeleev's classification, have the same structure as diamond and are semiconductors, with an energy band gap of 1.12 eV for silicon and 0.66 eV for germanium; tin is a metal.

Compounds made up of elements from columns 3 and 5, such as gallium arsenide GaAs, are semiconductors. Only the compounds AlSb and GaAs have forbidden energy bandwidths close to that of silicon and are suitable as component semiconductors. They have the same structure as diamond. Gallium arsenide has not been used as a computer material (the reasons for this are explained in the following chapters).

The "energy band gap" is the basic characteristic of semiconductors.

1.2.3. Intrinsic semiconductor: the electron–hole pair

The electrons in the outer layer of silicon and germanium form covalent bonds. Their energy is that of the valence band limit. If these electrons are given energy greater than the width of the energy band, for example, by thermal energy or by absorption of a photon (photoelectric effect (Volume 2, Chapter 4)), their (internal) energy becomes equal to or greater than that of the conduction band.

When an electron moves from the valence band to the conduction band, an electron is missing in the valence band: this is a "hole". An electron–hole pair is generated: a valence electron becomes a conductive electron which carries a negative charge and the hole in the valence band carries a positive charge (see section 1.2.5). A valence electron in a neighboring bond therefore has room to migrate from one hole to another. The hole moves in the opposite direction. The electrons in the conduction band and the holes in the valence band participate in electrical conduction.

	Width of the energy band gap E_G in eV	Intrinsic electron concentration at 25°C in atoms/cm $n_i = p_i$	Intrinsic resistivity at 25°C in ohm·cm	Electron mobility[7] in cm²/V·s at 25°C	Mobility of a hole µ in cm²/V·s at 25°C
Ge	0.67	2.9^{13}	46	3,600	1,800
Si	1.12	10^{10}	32,000	1,400	480
GaAs	1.43	2.25×10^6		8,000	400

Table 1.1. *Characteristic quantities of intrinsic semiconductors (Conwell 1952; Sze 2002, p. 36)*

In a "pure" material (in the theoretical absence of any impurity or dopant), at thermal equilibrium, the concentrations of intrinsic (conduction) electrons n_i and intrinsic holes p_i obey the law of mass action[8]:

$$n_i . p_i = n_i^2 = N_C . N_V . \exp \{-E_G / kT\} \text{ with } n_i = p_i \qquad [1.4]$$

with:

– E_G: the width of the energy gap;

– N_C: the density of states (number of electronic states per unit volume) of the conduction band;

– N_V: density of states (holes) in the valence band.

$N_C = 2.86 \times 10^{19}/cm^3$ and $N_v = 1.04 \times 10^{19}/cm^3$ for silicon at 300 K.

$N_C = 1 \times 10^{19}/cm^3$ and $N_v = 0.5 \times 10^{19}/cm^3$ for germanium at 300 K.

The concentrations $n_i = p_i$ of these intrinsic (electric charge) carriers at ordinary temperature are very low for silicon ($9.65 \times 10^9/cm^3$)[9] and much higher for germanium ($2.9 \times 10^{13}/cm^3$), as a result of a smaller energy bandgap than in silicon.

---

7 Speed of movement of an electron in an electric field of 1 V per centimeter.

8 The law of mass action is derived from the Fermi–Dirac distribution function (which reduces to the Boltzmann distribution), giving the probability of occupation of an energy level E by an electron at temperature T as a function of the "Fermi energy" $f(E) = \exp(-(E - E_F)/kT)$.

9 For copper, the electron density is identical to the atomic density: the number of copper atoms per cm³ is 8.5×10^{22}. For silicon, the number of atoms is $5 \times 10^{22}/cm^3$.

The exponential dependence of the concentration (number per unit volume) of "intrinsic charge carriers" on temperature, with the parameter "energy band gap" E_G, means that the concentration of electron–hole pairs increases with temperature according to this parameter. For silicon, it only reaches significant levels at high temperatures (Figure 1.9, dotted curve). For germanium, on the other hand, it reaches significant values (higher than the concentration of charge carriers provided by the dopants) at lower temperatures, resulting in unstable operation of germanium-based components at T > 75°C, whereas for silicon, stable operation extends up to 150°C.

If the width of the energy band gap is small, the intrinsic conductivity can be high even at low temperatures, and this is the conductivity measured on pure sulfides and oxides.

1.2.4. *Extrinsic semiconductor*

The discovery of the effect of doping of elements in the third column (boron) and the fifth column (phosphorus) on the conductivity of silicon, in 1940, by Russell Ohlof Bell Labs, is described in Chapter 2 (section 2.3.1.1).

In a semiconductor (Si or Ge) doped with phosphorus, a germanium (or silicon) atom is replaced by a phosphorus atom (five electrons, four of which saturate the crystal's bonding orbitals). The binding energy of the fifth electron to the phosphorus ion is close to the limit of the conduction band. All that is needed is a very low energy, the ionization energy $\Delta E = E_C - E_D$, for it to enter the conduction band and become a conduction electron (Figure 1.6). Conductivity of type n is provided by these conduction electrons. The same applies to the other donors Sb and As.

In a boron-doped semiconductor, an atom of germanium or silicon is replaced by an atom of boron (three electrons), leaving a hole occupying an energy level close to the valence band, and therefore easily falling back into the valence band (Figure 1.6). Conductivity of the p type is provided by valence electrons moving from hole to hole.

The doping donors n (of concentration N_D) are elements from column V (P, As, Sb).

The dopant acceptors p (concentration N_A) are elements from column III (B, Al, Ga).

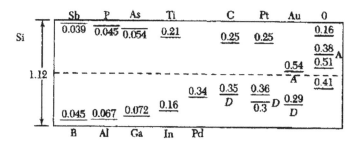

Figure 1.6. *Ionization energies (in eV) of various impurities (including n and p dopants) in silicon (Sze 2002, p. 38)*

Other elements confer extrinsic conductivity. Oxygen confers a conductivity of the type n to silicon (section 1.2.1).

At thermodynamic equilibrium, in an N semiconductor (n-doped), the concentrations of n conduction electrons $n_{(N)0}$ and p "holes" $p_{(N)0}$ obey the law of mass action (formula [1.5]):

$$n_{(N)0}.p_{(N)0} = n_i^2 = N_C.N_V. \exp\{-E_G/kT\} \qquad [1.5]$$

For an N (n-doped) (phosphorus) semiconductor, a fraction of the electrons introduced by the N_D doping "recombines" with the intrinsic holes. But this fraction is very small, all the smaller the width of the energy band gap E_G is large. We can therefore assume that the concentration of conduction electrons "$n_{(N)0}$" reached at equilibrium is practically equal to the concentration of donors N_D, and therefore, as a result of the law of mass action, the concentration of holes at equilibrium is:

$$n_{(N)0} = N_D \text{ and } p_{(N)0} = n_i^2/N_D \qquad [1.6]$$

For a P (p-doped) semiconductor (boron, aluminum), similarly, we can assume that the concentration of holes in the p valence band $p_{(P)0}$ is practically equal to the concentration of acceptors N_A, hence the concentration of electrons in the conduction band at equilibrium:

$$p_{(P)0} = N_A \text{ and } n_{(P)0} = n_i^2/N_A \qquad [1.7]$$

For N (n-type) silicon (n-doped), for N donor concentrations greater than $10^{14}/cm^3$, the concentration of holes (minority carriers) is very low. The same applies to P (p-type) silicon (p-doped), the concentration of (conduction) electrons (minority carriers) is very low.

For N (n-type) germanium (n-doped), on the other hand, for donor concentrations of $N_D > 10^{14}/cm^3$, the concentration of holes $p_{(N)0} = n_{i2} / N_D > 10^{12}/cm^3$ is relatively high.

1.2.4.1. *The Fermi energy (level)*

Fermi energy E_F of a conduction electron of an N semiconductor is its "free energy" also known as the "thermodynamic potential"[10]:

$$E_{F(n)} = E_C + kT \log (n_{(N)0}/N_C) = E_C + kT \log (N_D/N_C) \text{ with } E_{F(n)} < E_C \qquad [1.8]$$

with:

– $n_{(N)0}$: the equilibrium concentration of conduction electrons;

– N_C: the number of electronic states per unit volume (density of states) of the conduction band (see section 1.2.3).

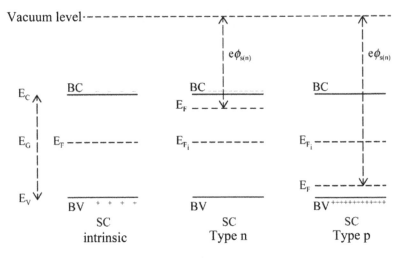

Figure 1.7. *Energy band diagram. Position of the Fermi level E_F in different types of doped semiconductors. Work function of an n or p dopant: $e\Phi_{s(n)}$, $e\Phi_{s(p)}$ (figure taken from Mathieu (1996, p. 60))*

10 Fermi energy is the free energy of a constituent of the system F = U – TS, where E is the internal energy of the component and S is its entropy. The thermodynamic potential of a constituent of the system (in this case a conduction electron: μ = dF/dn). This is the equivalent of the chemical potential of a component in a solution or the electrochemical potential of an ion in an electrolyte solution. See Mott and Jones (1958, p. 175), where it is clearly stated thermodynamic potential per electron: E_F = u – Ts + pv.

The Fermi energy determines the *work function* of a conducting electron in an N semiconductor (n-doped) $e\Phi_{sn}$: this is the energy required to extract a conduction electron from the semiconductor and bring it into a vacuum (Figure 1.7); the same applies to Fermi energy and work function of a hole in a P semiconductor.

Note that the *Fermi level* of a conduction electron is located in the band gap. It takes more energy to extract an electron from the conduction band and bring it into a vacuum than its internal energy. The Fermi level E_F depends on the doping in each type of semiconductor. The closer it is to the limit of the forbidden energy band (conduction or valence) band the higher the doping. Conversely, the lower the doping, the closer the Fermi level E_F is to the intrinsic Fermi level E_{Fi}.

Microelectronics components are made up of "blocks" (metal, N or P semiconductor) joined together along a plane (metal-semiconductor diode, PN junction diode, bipolar transistor with two junctions NP), with different Fermi levels. When the regions are brought into contact to make the component, because of the initial difference in Fermi levels, and in the absence of (electrical) polarization, electrons leave the region with the high Fermi level and move into the region with a lower Fermi level, until the thermodynamic potentials, that is, the "Fermi levels" of the electrons in the two regions in contact, are equal. In other words, at thermodynamic equilibrium, the Fermi levels of the electrons in the different regions in contact align (see Chapter 2, Figure 2.2 and Chapter 4, Figures 4.4 and 4.5).

1.2.5. Mechanism of generation/recombination of electron–hole pairs and "lifetime" of minority carriers

1.2.5.1. Mechanism of generation/recombination of electron–hole pairs

The "generation of an electron-hole pair", defined in section 1.2.3, is obtained by exciting an electron in the valence band, making it pass into the conductivity band, leaving a hole in the valence band, and vice versa for "electron-hole recombination". Both of these processes require or emit energy. They can occur either by absorbing a photon (as in the case of solar cells, Volume 2, Chapter 4), or by absorbing or emitting a phonon (quantum of energy supplied or absorbed by the vibrations of the network).

For silicon and germanium, electron-hole generation/recombination is termed "indirect", because it takes place in two stages, each of which involves only small amounts of energy (according to Shockley's mechanism–Read-Hall mechanism), via a generation/recombination center called a "trap", which occupies an intermediate energy level in the band of forbidden energy (localized levels). Impurity atoms (strictly speaking, excluding dopants) and lattice defects (grain boundaries and dislocations) constitute such "traps".

Figure 1.8 shows the two stages in the recombination or generation of an electron–hole pair:

– For electron–hole recombination step 1: capture of a conduction electron by a trap. The electron's energy falls to the energy level of the trap. The energy $E_c–E_t$ is transferred to the crystal lattice; step 2: the electron in the trap falls into the valence band and fills a hole.

– For electron–hole generation step 3: an electron from the valence band is transferred to a trap, leaving a hole in the valence band. The energy required for this step is $E_t–E_v$; step 4: the electron in the trap is transferred to the conduction band (the energy required $E_c–E_t$ must be supplied by light energy or by the thermal energy of the lattice).

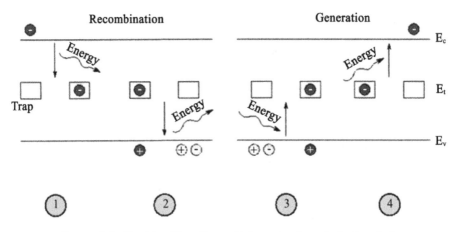

Figure 1.8. *The Shockley–Read–Hall mechanism of electron-hole generation/recombination in two steps via a trap (Entner 2007). For a color version of this figure, see www.iste.co.uk/vignes/silicon1.zip*

1.2.5.2. *Lifetime of minority carriers*

When a conduction electron has been "created" in a P semiconductor by electron-hole generation, or has been injected into a P semiconductor, during its movement it will recombine with a majority carrier (hole in a P semiconductor) after a certain time, losing its energy.

In a P semiconductor, the electron–hole recombination rate r is expressed by:

$$r = \Delta n / \tau_\pi = (n_{(P)} - n_{(P)0}) / \tau_{n(P)} \tag{1.9a}$$

where $\tau_{n(P)}$ is the minority carrier lifetime of the electron. This lifetime is the inverse of a first-order rate constant. Recombination only occurs for electrons in excess of the equilibrium concentration $n_{(P)} > n_{(P)0} = n_i^2 / N_A$ (expression [1.7]).

The same applies to a hole coming from the P region and injected into the N region. The recombination rate is expressed by:

$$r = \Delta p / \tau_{\pi(N)} = r = (p_{(N)} - p_{(N)0}) / \tau_{p(N)} \tag{1.9b}$$

The lifetime of the minority carriers is inversely proportional to the concentration of the generation/recombination centers of electron–hole pairs, the traps:

$$\tau_{n(P)} = 1 / N_T(P) \tag{1.10}$$

By minimizing traps, electron–hole recombination centers and therefore impurities and lattice defects, we increase the lifetime of minority carriers (which is what is sought in solar cells (Volume 2, Chapter 4)), and conversely, when short lifetimes are required, specific recombination centers are introduced into the semiconductor, such as gold (Figure 1.6) in bipolar transistors (Chapter 5, section 5.1.3.2).

1.2.6. *Influence of temperature on the concentration of majority carriers*

Figure 1.9 shows the variation of the concentration of (conduction) electrons (majority carriers) with temperature in an N semiconductor (n type) for a concentration of donors (in this case electrons) of $N_D = 10^{15}$ atoms/cm^3 in silicon.

There are three temperature ranges. At very low temperatures, the freeze-out region, the concentration of intrinsic electrons is almost zero, the concentration of the majority carriers (conduction electrons) of the dopant in the silicon N (n-type) increases with temperature[11]. Between 100 and 500 K, known as the extrinsic region (depletion regime), the effective concentration of majority carriers is equal to the concentration of donors (expression [1.6]).

---

11 The curve showing the variation in the resistivity of silicon as a function of temperature (Figure 1.2), at very low temperatures, corresponds to this first region, and all microelectronic components operate outside this range.

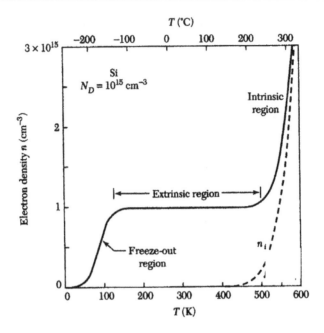

Figure 1.9. *Variation with temperature of the concentration of conduction and intrinsic electrons (dotted curve) in N (n-type) silicon (Sze 2002, p. 43)*

Above 500 K, in the intrinsic region, the concentration of intrinsic carriers dominates and increases strongly with temperature. The temperature above which the intrinsic regime dominates is much lower for germanium (75°C), whose ionization energy (0.67 V) is much lower than that of silicon (1.12 eV) (Table 1.1).

1.2.7. *Conduction current in a semiconductor*

In a bar of N (n-type) semiconductor (n-doped) polarized by the application of a potential between its ends, under the action of the electric field thus created, the conduction electrons move freely toward the positive pole and the valence electrons move from hole to hole also toward the positive pole. The holes move toward the negative pole. The conduction current (drift current) is the sum of an electron current and a current of holes flowing in opposite directions (Figure 1.10(a)).

To supply the current of electrons, electrons are "injected" at the negative electrode, a fraction circulates in the material as conduction electrons and is extracted by the positive electrode. The other fraction falls into the holes and becomes valence electrons, moving from hole to hole until it reaches the positive electrode, where it is freed from its covalent bond and becomes a conduction electron again, to be extracted by the positive electrode, leaving a hole. These holes move in the opposite direction to the negative electrode, recombining with electrons injected at this electrode. In the extrinsic domain (Figure 1.9), in an N semiconductor, the current of holes is very low and vice versa for a P semiconductor.

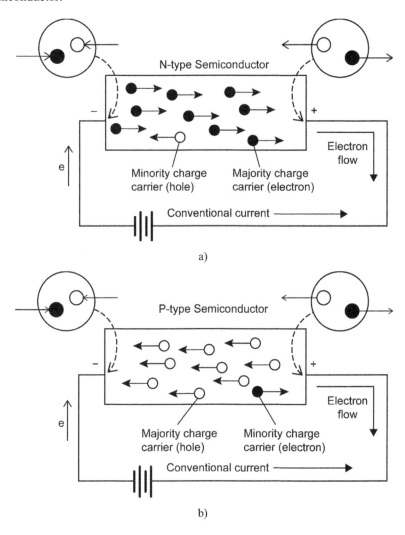

a)

b)

Figure 1.10. *Electron and hole currents in (a) an N semiconductor and (b) a P semiconductor. The metal/semiconductor contacts are ohmic (adapted from Wells 2009)*

From Ohm's law $V/I = R = \rho l/S$, we define the current density per unit area J, which is expressed as a function of the electric field $E = V/l$ by:

$$J = I/S = \sigma E = E/\rho \qquad [1.11]$$

where σ is the conductivity and ρ is the resistivity (ohm.cm).

The conductivity of a semiconductor is the sum of the products of the concentration (by volume) of electrons (conduction electrons) and holes (valence electrons) multiplied by their respective mobilities:

$$\sigma_N = e \cdot \mu_e \cdot n + e \cdot \mu_p \cdot p \qquad [1.12]$$

where n is the concentration of (conduction) electrons, the majority carriers, with mobility μ_n, p is the concentration of holes, minority carriers, with mobility μ_p in a semiconductor N.

The conductivity σ_N of a semiconductor N, in the extrinsic domain, is provided mainly by the electrons supplied by the donor $n = N_D$.

The conductivity σ_P of a p-type semiconductor is provided mainly by the holes created by the N dopant ($p = N_A$).

1.2.8. *Mobility of majority carriers: electrons and holes*

The mobility of the majority carriers in an extrinsic semiconductor varies with temperature (Figure 1.11) and with doping (Figure 1.12).

At a given temperature, the mobility of electrons and holes varies according to the concentration of the respective dopant. But up to relatively high concentrations, 10^{16} atoms/cm^3, it is more or less constant (Figure 1.12).

In a semiconductor, the movement of a conduction electron in the crystal is disordered, colliding with the atoms, or more precisely undergoing repulsive forces (as a result of coulombic interactions) with the atoms of the crystal vibrating under the action of thermal energy (lattice scattering), and with the ions of the dopants (attractive forces) (ionized impurity scattering). The same applies to holes.

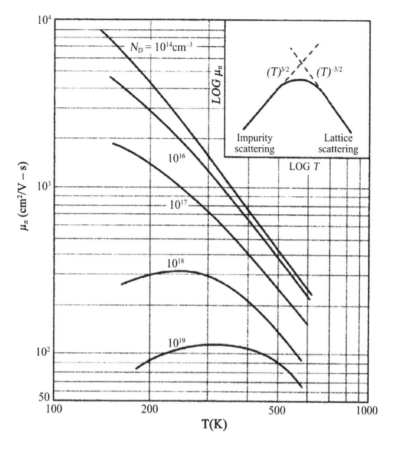

Figure 1.11. *Electron mobility in silicon N as a function temperature and for different donor of $N_D = n_{(N)}$ (Sze 2002, p. 50)*

In an electric field, the overall movement of an electron takes place in the opposite direction to the field, but remains disordered. Electrons in the conduction band and holes in the valence band very quickly acquire a constant speed.

The mobility of electrons in an N semiconductor and holes in a P semiconductor due to interactions with lattice atoms, lattice scattering, varies as a function of temperature according to a $T^{-3/2}$ law. It increases as the temperature decreases.

The mobility of electrons due to interactions with donor and acceptor ions (ionized impurity scattering) varies according to a $T^{3/2}$ law. It increases as the temperature rises.

The mobility of electrons and holes in germanium is three to four times greater than in silicon at ordinary temperature (see Table 1.1), a characteristic that was a positive factor in the choice of germanium for components such as diodes acting as switches in early computers.

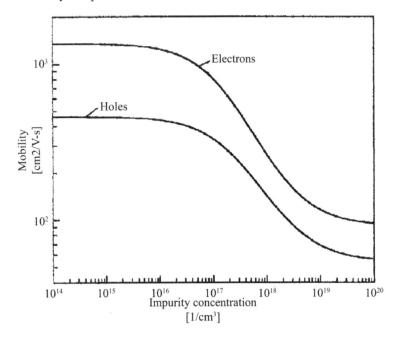

Figure 1.12. *Electron and hole mobility in silicon, at T = 300 K as a function of their concentrations (atoms/cm³) (Gray 2003)*

1.2.9. *Resistivity of silicon and germanium*

In the temperature range where the effective number of majority carriers participating in conduction is that of the donor or acceptor, the "extrinsic region" (Figure 1.9), at a given temperature, conductivity varies as a function of dopant concentration (Figure 1.13 for silicon and Figure 1.14 for germanium). For light doping, corresponding to a dopant concentration of 10^{15} atoms/cm³, the resistivity of silicon is 6–10 ohm.cm, while the resistivity of germanium is of the order of 20–40 ohm.cm. The important feature is that the resistivity varies greatly with the concentration of the dopant, which gives great control over the conductivity/resistivity. Conductivity/resistivity can be modulated over more than 10 orders of magnitude. This is the major attraction of Si and Ge semiconductors.

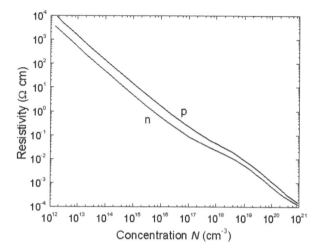

Figure 1.13. *Variation in the resistivity of monocrystalline silicon with the concentration of n and p dopants at ordinary temperature (IOFFE-Si). For a color version of this figure, see www.iste.co.uk/vignes/silicon1.zip*

For silicon, light doping corresponds to a concentration of dopants between 10^{14} and 10^{16}/cm³ (n⁻ or p⁻); normal doping from 10^{16} to 10^{19} /cm³ (n or p) and heavy doping above 10^{19}/cm³ (n⁺ or p⁺). A boron doping of 4.9×10^{13} atoms/cm³ (1 ppba or 0.38 ppbwt) corresponds to one B atom per billion silicon atoms. For P, it is 7.2×10^{13} a/cm³ (1.44 ppba = 1.59 ppbwt). An important aspect of this purity is the distance between two atoms (ions) of dopant (1 μm). This implies that a dopant ion can be considered totally isolated and without any interaction with other dopant ions.

In the extrinsic region (Figure 1.9), for a given concentration of dopant, the variation of conductivity as a function of temperature is due to the variation in mobility with temperature (Figure 1.11). For dopant concentrations of less than 10^{18} atoms/cm³ (i.e. relatively high), conductivity decreases with increasing temperature, so the resistivity of an N semiconductor in the extrinsic domain increases with temperature, as it does for metals. This explains why (impure) silicon has been classified as a metal.

Above a certain temperature (Figure 1.9), in the "intrinsic domain", the concentration of intrinsic free carriers becomes greater than that of the dopant; conductivity then increases rapidly with temperature (the concentration of intrinsic carriers increases with temperature). It increases much more with temperature for germanium, whose energy band gap (0.67 eV) is much smaller than that of silicon (1.12 eV). In the case of germanium, "unstable" operation (above the extrinsic range) of the components appears as early as 75°C. Silicon operates "stably" at higher temperatures up to 200°C.

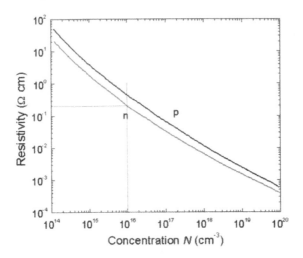

Figure 1.14. *Variation in the resistivity of germanium with dopant concentration (IOFFE-Ge). For a color version of this figure, see www.iste.co.uk/vignes/silicon1.zip*

This was the main factor that led the US military in the 1950s to impose silicon as the basic material for electronic components used in missiles, etc.

1.3. References

Becket, F.M. (1921). Process of refining crude electric-furnace silicon. Patent, US1386227A.

Benedicks, C. (1915). La conductivité du germanium était intermédiaire entre celle du silicium et celle de l'étain. *Int. Z. Metallographie*, 7, 225–238.

Berzelius, J.J. (1824). Décomposition du fluate de silice par le potassium. *Annales de chimie et de physique*, 27, 337–359.

Bidwell, C.C. (1922). Resistance and thermo-electric power of metallic germanium. *Physical Review*, 19, 447.

BRGM (2019). Mineral info [Online]. Available at: https://www.minrealinfo.fr.

Brockbank, C.J. (1916). Process for purifying silicon. Patent, US1180968.

Burdette, S.C. and Thornton, B.F. (2018). The germination of germanium. *Nature Chemistry*, 10, 244.

Busch, G. (1989). Early history of the physics and chemistry of semiconductors – From doubts to fact in a hundred years. *European Journal of Physics*, 10(4), 254–264.

Conwell, E.M. (1952). Properties of silicon and germanium. *Proceedings of the IRE*, 40(10), 1327–1337 [Online]. Available at: www.ioffe.ru/SVA/NSW/semiconductor.

Davy, H. (1808). Electrochemical researches on the decomposition of earths with observations on the metals obtained from the alkaline earths and on the amalgam procured from ammonia. *Philosophical Transactions of the Royal Society of London*, 98, 333–370.

Deville, H. (1854). Note sur deux procédés de préparation de l'aluminium et sur une nouvelle forme de silicium. *Comptes-rendus de l'académie des sciences*, 39(3), 321–326.

Deville, H. (1855a). Du silicium et du titane. *Comptes-rendus de l'académie des sciences*, 40, 1034–1036.

Deville, H. (1855b). Recherches sur les métaux et en particulier sur l'aluminium et sur une nouvelle forme de silicium. *Annales de chimie et physique*, 43, 5–31.

Entner, R. (2007). Modeling and simulation of negative bias temperature instability. PhD Thesis, Vienna University of Technology, Vienna [Online]. Available at: www.iue. tuwien.ac.at/phd/entner.

Faraday, M. (1833). Experimental researches in electricity. On a new law of electric conduction. *Philosophical Transactions of the Royal Society*, 123, 507–522.

Gay-Lussac, J.L. and Thénard, L.J. (1811). *Recherches physico-chimiques, faites sur la pile ; sur la préparation chimique et les propriétés du potassium et du sodium ; sur les acides fluorique, muriatique et muriatique oxigéné ; sur l'action chimique de la lumière ; sur l'analyse végétale et animale, etc.* Déterville, Paris.

Gray, J.L. (2003). The physics of solar cell. In *Handbook of Photovoltaic Science and Engineering*, Luque, A. and Hegedus, S. (eds). Wiley, New York.

Hittorf, J.W. (1851). Ueber das elektrische leitungvermögen des schwefelsibers. *Ann. Phys. Lpz.*, 84, 1–28.

IOFFE (n.d.). Si – Electrical properties. Resistivity versus impurity concentration for Si at 300 K [Online]. Available at: http://www.ioffe.ru/SVA/NSM/Semicond/Si.

IOFFE (n.d.). Ge – Electrical properties. Resistivity versus impurity concentration for GE at 300 K [Online]. Available at: http://www.ioffe.ru/SVA/NSM/Semicond/Ge.

Johnson, W.C., Foster, L.S., Krauss, C.A. (1935). The extraction of germanium and gallium from germanite. The removal of germanium by the distillation of germanous sulfide. *J. Am. Chem. Soc.*, 57(10), 1828–1831.

Koenigsberger, J. and Schilling, K. (1908). Uber die Leitfähigkeit einiger fester Substanzen. *Physikalische Zeitschrift*, 9, 347–352.

Lavoisier, A. (1789). *Traité élémentaire de chimie, présenté dans un ordre nouveau et d'après les découvertes modernes*, volumes I and II. Cuchet, Paris.

Lavoisier, A. (1965). *Traité élémentaire de chimie, présenté dans un ordre nouveau et d'après les découvertes modernes*, volumes I and II. Reprint, Cultures et Civilisations, Brussels.

Mathieu, H. (1996). *Physique des semiconducteurs et des composants électroniques*, 3rd edition. Masson, Paris.

Meissner, W. and Voigt, B. (1930). Messungen mit hilfe von flüssigen Helium IX, Wider-stand der reinen Metalle in Tiefen temperature. *Annalen der Physik*, 399(8), 892–936.

Moissan, H. (1897). *Le four électrique*. Gallica, Paris.

Mott, N.F. and Jones, H. (1958). *Metals and Alloys*. Dover Publications, New York.

Patnode, W.I. and Work, R.W. (1931). Extraction of germanium and gallium from germanite. *Ind. Eng. Chemistry*, 23(2), 204–207.

Pearson, G.L. and Brattain, W.H. (1955). History of semiconductor research. *Proceedings of the IRE*, 43(12), 794–806.

Scaff, J.H. (1946). Preparation of silicon materials. Patent, US2402582A.

Seitz, F. and Einspruch, N.G. (1998). *Electronic Genie; The Tangled History of Silicon*. University of Illinois Press, Illinois.

Sze, S.M. (2002). *Semiconductor Devices*, 2nd edition. Wiley, New York.

Teal, G.K. and Kraus, C.A. (1950). Compounds of germanium and hydrogen. *Jour. Amer. Chem. Soc.*, 7, 4706–4709.

Tone, F.J. (1913). Process for producing silicon. Patent, US906338.

Tücker, N.P. (1927). *Journal of Iron and Steel Institute*, 115, 412–416.

Vignes, A. (2011). *Extractive Metallurgy: Basic Thermodynamics and Kinetics*. ISTE Ltd, London, and John Wiley & Sons, New York.

Warren, H.N. (1888). Detection and estimation of selenium in meteoric iron. *Chem. News*, 57, 54.

Wells, C.J. (2009). The diode [Online]. Available at: https://www.technologyuk.net/science/electrical-principles/the-diode.shtml.

Wilson, A. (1931). The theory of electronic semiconductors. *Proc. Royal Society*, 133, 458.

Winkler, C. (1886). Mittheilungen über des Germanium. Zweite Abhandlung. *J. Prak. Chemie*, 34, 177–209.

Wöhler, F. (1857). Uber das aluminium. *Annales de Poggendorff*, 11, 146–161.

2

The Point-Contact Diode

The point contact diode consists of a semiconductor wafer, a metallic tip (whisker) (W, Pt, Au) applied to one side of the wafer and a second flat electrode applied to the other side of the wafer.

Following the discovery, in 1874, of the current rectifier effect of the point contact diode and, in 1901, the discovery of the detection of "radio waves" by the galena-based point contact diode, this diode became the component of the AM radio wave reception circuit in galena receivers. The purely empirical search for materials for the point contact diode led to the emergence of silicon in 1906.

Research into the radar detection of enemy aircraft during the Second World War led to the re-emergence of the silicon point contact diode as the basic "frequency converter" component of radars, and to the development of the germanium point-contact diode.

This research led to the major discovery of silicon N and silicon P, in 1940; in other words, the discovery that the conductivity of silicon, depending on the doping with elements such as phosphorus or boron, could be of the n type (by electrons) or p type (by holes) – a property also observed for germanium.

The N and P silicon-based point-contact diodes have the rectifier effect, while only the germanium N diode has the rectifier effect.

Controlling the conductivity of silicon and germanium leads to metallurgical developments (purification, heat treatment).

The germanium N point contact diode replaced diodes (vacuum tubes) in consumer and specialist telecommunications applications from the end of the Second World War until the end of the 1950s.

Prototype computers, built in 1949 and 1950, used germanium N point-contact diodes in logic circuits because of their high switching speed.

This chapter presents:

– the point-contact diode and its functions. The basics of its operation and functions;

   – the history of:

      - the discovery of the rectifier effect of the point-contact diode,

      - the discovery of the detection of radio waves by the point-contact diode,

      - the silicon-based point-contact diode,

      - the germanium N point-contact diode,

      - the point-contact diode as frequency converter in radar receivers during the Second World War;

   – the discovery of silicon N and silicon P;

   – the metallurgical developments (purification, heat treatment) of the two materials: silicon and germanium;

   – the industrial development of the germanium-based diode after the Second World War;

   – in the Appendix: currents in a metal-semiconductor diode, physical bases: the case of the germanium N point-contact diode.

2.1. Features and functions

2.1.1. *Characteristic curve, rectifier effect*

The diode is a component that allows electric current to flow in only one direction. The vacuum diode is made up of an emissive cathode and an anode placed in a glass enclosure; the solid diode is a component made up of two joined regions: a metal and a semiconductor: Schottky diode, where the contact is planar, and point-contact diode, where the contact is punctual; PN diode between two semiconductors (presented in Chapter 4).

The point-contact diode consists of an N (n-doped) or P (p-doped) semiconductor wafer, a metallic tip (whisker) (W, Pt, Au) applied to one side of the wafer to make a "rectifying contact" and a second flat electrode applied to the other side of the wafer to make an "ohmic contact" (without resistance). For some diodes (germanium-based), the contact is made by soldering between the metal tip and the chip.

This component, known as a "rectifier", only allows current to flow in one direction, the "easy direction of current flow".

Figure 2.1 shows a typical DC current–voltage characteristic curve for a silicon diode exhibiting the rectifier effect. Note that the current in the "on" direction is low, from 15 to 20 mA, for a voltage of 1 V. In the "blocked" (or "reverse") direction, the resistance is very high. A very small current flows, which is known as the saturation current.

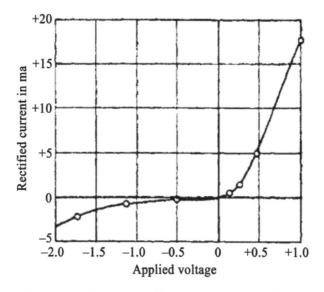

Figure 2.1. *Characteristic curve (current–voltage) of a silicon diode (Torrey and Whitmer 1948, p. 20)*

Above a certain negative voltage, known as the *breakdown voltage* (breakdown potential, peak reverse voltage), the current becomes very high and destroys the diode. The electric field generated by the high voltage produces an avalanche of electrons that destroys the diode.

2.1.2. *Rectifier contact and ohmic contact*

To explain the rectifying effect presented by a metal-semiconductor contact (plane contact or point contact), we need to consider the energy band diagram before contact between the two components, and after contact, when equilibrium is reached (Sze 2002, p. 225).

When a metal m and a semiconductor N (doped n) are brought into contact, for Fermi energies of the electrons of the metal E_{Fm} or $e\Phi_m$ and of the semiconductor $E_{Fs(n)}$ (the Fermi energy of a semiconductor is defined and presented in Chapter 1, section 1.2.4, Figure 1.7), such that:

$$E_{Fm} < E_{Fs(n)} \text{ or } e\Phi_m > e\Phi_{s(n)} \text{ (work function)}$$

the potential barrier. Φ_{Bn}:

$$\Phi_{Bn} = (\Phi m - \chi)$$

opposes the passage of electrons from the metal into the semiconductor (Figure 2.2(a)).

On the other hand, electrons from the N semiconductor, with Fermi energy $E_{Fs(n)} > E_{Fm}$, pass into the metal, leaving fixed positive charges in the region of the semiconductor in contact with the metal (the positive ions of the dopant: the donor). The passage of electrons from the semiconductor into the metal continues until the Fermi levels are aligned, with the formation of a zone (metal tip) or layer (plane contact) progressively depleted in electrons, called the depletion zone or layer (also known as a space charge zone or layer). This leads to the formation of a potential barrier V_{bi} (built-in potential), which prevents the continued passage of electrons from the semiconductor into the metal (Figure 2.2(b)):

$$V_{bi(n)} = \Phi_m - \Phi_{s(n)} > 0 \qquad\qquad\qquad [2.1]$$

In direct polarization, at a positive potential $V_F > 0$, the potential barrier is lowered ($V_{bi(n)} - V_F$), the width of the depletion zone (layer) decreases and a current of electrons flows in the N semiconductor direction$\rightarrow$ metal (Figure 2.11b). In reverse polarization, the potential barrier is increased ($V_{bi(n)} + V_R$), the width of the depletion zone (layer) increases and a very small current flows.

Conversely, if the work functions are such that $\Phi_m < \Phi_{s(n)}$, no potential barrier is formed. Electrons can pass from the metal into the semiconductor, and the contact offers only a slight resistance to the flow of current: the contact is ohmic.

Similarly, when a metal or a metal tip and a semiconductor P (p-doped) are brought into contact, for $e\Phi_m < e\Phi_{s(p)}$, electrons pass from the metal into the semiconductor and recombine with the holes (Figure 2.11a), creating a depletion zone (layer) made up of negative charges (the negative ions of the dopant: acceptor) and a potential barrier is formed, which progressively prevents the passage of electrons from the metal into the semiconductor:

– $V_{bi(p)} = \Phi_{s(p)} - \Phi_m > 0$, the contact is rectifying;

– conversely, if $e\Phi_m > e\Phi_{s(p)}$, the contact is ohmic.

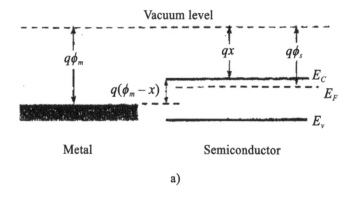

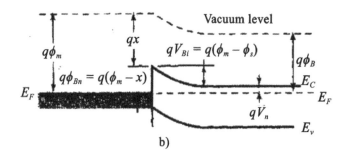

Figure 2.2. *Semiconductor-metal (W/Si-N) diode energy band diagram and potential barriers $q\Phi_{Bn}$ and qV_{bi} : (a) before contact; (b) after contact and equilibrium (Sze 2002, p. 226)*

Silicon (N- or P-doped) chips and Al, Au or W tips exhibit the rectifier effect. Only germanium N-based diodes have the rectifier effect (section 2.3.2.2).

2.1.2.1. *Ohmic contact by tunnel effect*

Ohmic contact, even with a rectifying couple such as $\Phi_m > \Phi_s$ (Al)/Si(n), can be achieved by a high doping of the semiconductor, $N_D > 10^{19}$ /cm^3; the thickness of the barrier layer becomes very low and the electrons cross the interface as a result of the tunnel effect. In transistors and integrated circuits, ohmic contacts are made using the tunnel effect (see Chapters 5 and 7).

2.1.3. *Point contact diode functions*

2.1.3.1. *Component of an AM radio wave reception circuit*

The detection of AM radio waves[1] was the first use of such diodes, between 1900 and 1920, in radio receivers known as "cat's whisker crystal radio receivers".

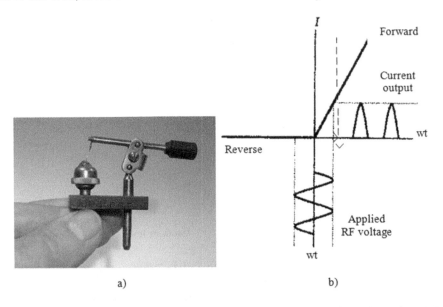

a)   b)

Figure 2.3. *(a) Point-contact diode with a galena crystal from an antique Swedish radio used as early as 1906 (Hellgaard); (b) idealized characteristic and rectification of an alternating signal by a point-contact diode. For a color version of this figure, see www.iste.co.uk/vignes/silicon1.zip*

The diode rectifies the low-frequency electrical signal generated by the AM radio wave reception circuit. The principle of rectifying an alternating (low-frequency) current is shown in Figure 2.3(b).

Figure 2.4(a) shows the circuit of an AM radio set consisting of an AM radio wave reception circuit (antenna + L1-C1 oscillating circuit) used to select a frequency (b), a point-contact diode acting as a signal rectifier (c), a capacitor C2, which extracts the high-frequency component of the signal, and the low-frequency signal (sound component) (d) being fed into headphones with a high resistivity (1–2 kΩ).

---

1 Detection: this is the name given to the operation which consists of producing, from a sinusoidal voltage, a DC voltage proportional to its amplitude (detection of amplitude modulated signals). This term is synonymous with rectification. It is reserved for cases where the powers involved are low.

Such a circuit made it possible to listen to AM radio waves as early as 1901. These diodes were used until 1920, when they were replaced by vacuum diodes. They had a second life during the Second World War and into the 1950s (section 2.4).

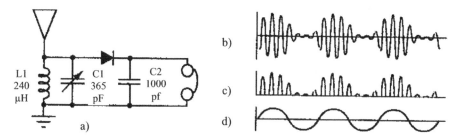

Figure 2.4. *(a) Circuit of an AM radio wave receiver with a point-contact diode; (b) signal supplied by the antenna plus oscillating circuit; (c) signal rectified by the diode; (d) audio signal received by the headphones*

2.1.3.2. Component of a "frequency converter"

The point-contact diode allows the high-frequency currents generated by circuits receiving high-frequency electromagnetic waves – microwaves – to pass through. It no longer functions as a current rectifier. It is used as a frequency converter in superheterodyne receivers for radio, television and satellite communications, and was developed and used extensively during the Second World War in radar receivers (Figure 2.5).

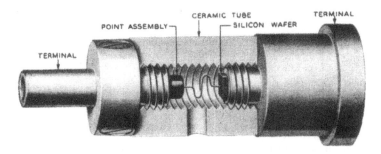

Figure 2.5. *Point-contact diode: mixer of a frequency converter for radar reception (Scaff and Ohl 1947, p. 10)*

The vacuum tubes previously used in radar receivers only allowed the reception of relatively long wavelengths, which seriously limited the spatial resolution of the resulting images. Hence, there is the need to use wavelengths in the microwave range (wavelength < 10 cm; f > 3 GHz).

NOTE.– In a vacuum diode, the electrons take a certain time (transit time t_t) to cross the distance (1 mm) between the cathode and the anode, which leads to a maximum operating frequency: fmax = $1/t_t$, of the order of 108 Hz, that is, 30–100 times too low for a vacuum diode to be usable. However, in the case of a point-contact diode, the distance that the electrons have to cross (the thickness of the "depletion zone" (see Figure 2.2)) typically measures less than 1 μm and is therefore 4 orders of magnitude smaller than the distance between the cathode and anode of a vacuum tube, allowing currents with frequencies >3 GHz to pass through.

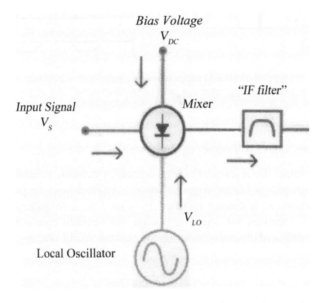

Figure 2.6. *Frequency converter circuit. For a color version of this figure, see www.iste.co.uk/vignes/silicon1.zip*

The point-contact diode allows high-frequency currents from the radar wave receiver circuit to pass through, hence its use as a "frequency converter". The diode transforms the high-frequency electrical signal (voltage) (input signal) from the receiver circuit into a lower frequency signal of 30 MHz/s that can be rectified and amplified. The basic circuit of a frequency converter consists of a diode and a filter. The "mixer" diode receives two electrical signals of similar frequencies, the frequency of the FM or radar radio signal and the frequency of a local oscillator. The signals are "multiplied". The diode's output signals contain the two basic HF frequencies, a high frequency and a low frequency IF: $f^{IF} = f^{HF} - f^{local}$. The filter short-circuits the HF signals. The mixer therefore shifts the received signal to a lower frequency. This converts a high-frequency signal of 10–100 GHz into a low-frequency signal of 30 MHz/s, which can be rectified and amplified.

Experimentally, it was found that only boron-doped silicon diodes P were effective as mixers for radar frequencies. These were the ones used exclusively during the Second World War.

2.1.3.3. Germanium point-contact diodes: switching components for logic circuits (switches)

The germanium N diode was used from the 1950s onwards as a component of logic circuits (switches) in computers because of its high switching speed.

2.1.4. Physical basis for the operation of a point contact diode

2.1.4.1. Frequency behavior[2]

The diode (metal-semiconductor) rectifies low-frequency currents. It allows high-frequency currents to pass through. The Schottky diode has the same characteristic. This characteristic is due to the formation of a depletion zone (layer) (called barrier region in Figure 2.7) (section 2.1.2 and Figure 2.2) in the semiconductor in contact with the metal, which behaves like a resistor for low-frequency currents and like a capacitor allowing high-frequency currents to pass up to a certain frequency: the cut-off frequency. The point contact diode allows higher frequency currents to pass than the Schottky diode, which is why it is used in radar.

The equivalent electrical circuit of the point-contact diode (Figure 2.7) consists of a capacity C, a parallel resistance R and an ohmic resistance r (Torrey and Whitmer 1948, p. 97).

The depletion zone (cat's whisker), depopulated of charge carriers (electrons for an N semiconductor, holes for a P semiconductor), constitutes a "capacity" C whose charge Q is a function of the potential applied to the diode:

$$C_J = Q/V = \varepsilon\varepsilon_0 \, A/e$$

where A is the contact area, e is the thickness of the barrier layer (of the order of a micron), which decreases as the forward voltage increases, and ε is the dielectric permittivity. The capacity of the depletion zone is very low (0.2–2.0 pF), due to the very small contact area (diameter of the order of 10^{-3} to 10^{-4} cm). The higher the frequency ω the signals, the lower the impedance $1/jC\omega$. Une capacity C of 1 pF has an impedance of 50 Ω.

---

2 A complete description of the frequency behavior of the point contact diode can be found in Torrey and Whitmer (1948). The theory of diode behavior was formulated by Hans Bethe (1942), the future Nobel Prize winner, best known for his participation in the Manhattan Project.

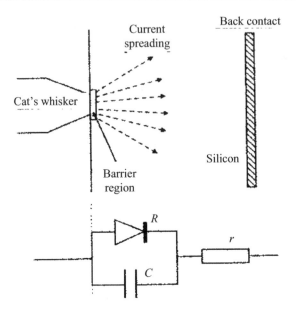

Figure 2.7. *Equivalent circuit and physical representation of the components of a point-contact diode (Orton 2009, p. 40)*

The resistance R of the depletion zone (barrier region) is high in the forward direction at low potentials, as long as there is a depletion zone; very high (of the order of 5,000 Ω)) in the opposite direction ("blocked") as a result of the increase in thickness of this depletion zone as the negative potential increases.

The ohmic resistance r (spreading resistance) of the semiconductor wafer is inversely proportional to the conductivity of the semiconductor σ (i.e. the concentration of majority carriers) and to a, the radius of the contact area A between the metal tip and the semiconductor block (very small):

$$r = 1/4\, \sigma\, a$$

The resistance r is high: 20–40 Ω because of the small contact radius.

The current with frequency $f = \omega/2\pi$ flowing through the capacitor is:

$$\overline{I}_C = (1/Z_C)\, V = C\omega V$$

and the current through the resistor R is:

$$\overline{I}_R = V/R$$

At low frequencies $\omega CR \ll 1$, capacity plays no part. The diode has a rectifying effect due to the resistance of the "barrier layer" R. The diode only conducts during positive voltage alternations, giving a pulsed rectified current (Figure 2.3(b)).

The higher the frequency, the higher the fraction of current flowing through the capacitor C. The capacitor then acts as a shunt, short-circuiting the resistance of the barrier region R. The diode is therefore equivalent to a circuit made up of the capacitor and the resistance r, with a time constant $\tau_{rC} = r.C$. The diode allows high-frequency currents to flow up to the diode's "cut-off frequency" of the diode:

$$f_c = 1/2\pi \, r.C$$

that is, frequencies from 1 to 10 GHz ($\lambda = 3$ cm). The diode is then used as the "mixer" of a frequency converter in radars.

2.1.4.2. Frequency converter mixer

The basic circuit of a frequency converter consists of a diode and a filter (Figure 2.4(b)). The "mixer" diode receives two electrical signals of neighboring frequencies, the frequency of the FM or radar radio signal and the frequency of a local oscillator; it then generates new frequencies. The mixer function is due to the nonlinearity of the I-V current-voltage characteristic of a point-contact diode (Figure 2.1) at low voltages (expression [2.2] in Appendix 2.5).

For a linear component (linear characteristic curve), the output frequencies are identical to those injected at the input; whereas, for a nonlinear component, the superposition of the two frequencies v_{signal} and V_{local} generates two new frequencies:

$$(f_{HF} + f_{local}) \text{ and } (f_{HF} - f)_{local}$$

The nonlinear characteristic of the point contact diode can be approximated by a law of the type:

$$I = a \, V^2 \text{ (square law assumption)}$$

The voltage across the diode is the sum of the voltages (Figure 2.4(b)):

$$V = V_{DC} + Vs + V_{LO}$$

The current through the diode is:

$$I(t) = a \, (V_{DC} + A \sin\{2\pi f_S \, t\} + B\cos\{2\pi \, f_l.t\})^2$$

By making this expression explicit, we show that the diode "converts" the sum of the two sinusoidal currents of high-frequency signals injected into the diode: the frequency of the radar signal and a local oscillator of neighboring frequency into signals of frequencies:

$$2f_{HF}, 2f_{local}, (f_{HF} + f_{local,}) \text{ and } (f_{HF} - f_{local})$$

including a high-frequency signal $f^*_{HF} = f_{HF} + f_{local}$ and a low-frequency signal: $f_{BF} = f_{HF} - f_{local}$.

The filter short-circuits HF signals. The circuit therefore shifts the received signal to a lower frequency. This converts a high-frequency signal of 10–100 GHz into a low-frequency signal of 30 MHz/s that can be rectified and amplified.

2.1.4.3. Switching

The point-contact diode was used in the logic circuits of the first "computers" because of its high switching speed. The switching time is the time taken to store or empty the charge Q of the capacity constituted by the depletion region. It is therefore very short, resulting in a high switching speed.. The storage time is in the order of 3 τ_{rC}, where τ_{RC} is the circuit time constant r C.

2.2. History

2.2.1. Discovery of the "rectifier effect"

In 1874, Karl Ferdinand Braun (Nobel Prize in Physics in 1909), like many of his contemporaries, was interested in the conduction of electricity in crystals of metal sulfides, such as galena and pyrite, which conduct electric current without electrolysis; he found that these crystals do not seem to have the same conductivity depending on the direction of the current flowing through them and that they do not therefore follow Ohm's law, "particularly when an electrode made contact over a small surface area" (Braun 1909). He had discovered the rectifying effect that these crystals confer on a pointed diode (Braun 1874). However, the deviation from Ohm's law was small (Georges and Mauviard 1997).

2.2.2. Discovery of AM radio wave "detection" by the point-contact diode

History credits Jagadis Chandra Bose in 1901 for the discovery of point-contact detection (Emerson 1998). In January 1897, Bose demonstrated his apparatus for generating and receiving radio waves at the Royal Institution in London.

Detection of the signal induced in the receiving antenna is achieved by a component consisting of a sharp iron tip pressed onto an iron surface. In 1901, Bose filed a patent, which was granted in 1904. Patents were granted for a device making up the point detector and various detection materials, including galena (PbS) (Bose 1904). This patent triggered a search for other crystals with rectifying properties equal to or better than galena.

After inventing the cathode-ray tube in 1898, Ferdinand Braun began work in the field of wireless telegraphy, at the request of a group of German companies that would become the "Telefunken Company", who were anxious to develop wireless telegraphy without using Marconi's patents. Barely a year later, he discovered a transmission system (which he patented), the "tuned circuit", enabling radio communications to be extended over longer distances.

When radio waves were first used to transmit spoken messages, Ferdinand Braun, having established that the rectifier effect of the point-contact diode persisted at low frequencies, proposed the use of this type of detector to the Gesellschaft für Drahtlose Telegraphie (Wireless Telegraphy Co) in 1901, which finally adopted it in 1905.

2.2.3. *Discovery of the silicon point-contact diode*

Greenleaf Pickard is credited with the discovery in 1906 of the rectifying effect of a point-contact diode made from a silicon wafer: "a landmark year" (Seitz and Einspruch 1998, p. 29). Between 1902 and 1906, Pickard, a researcher at ATT (the parent company of Bell Labs), tested 250 "minerals" and electric furnace products. Among them, metallurgical silicon obtained from Westinghouse Electric Corporation[3] proved to be one of the best detectors, along with galena, molybdenite and zinc oxide (perikon). Pickard took out numerous patents (Figure 2.6) (Pickard 1906a) and presented his discoveries, including improvements to the detection assembly, in various publications (Pickard 1906b, 1909, 1919), in which he explained the reasons for his choice: Pickard thought that the detection of the signal, induced by the reception of radio waves, by contact between different conductors, was due to the thermoelectric effect. Therefore, the best results should

---

3 The first use of metallurgical silicon produced in electric furnaces since 1995 (see Chapter 1, section 1.1.3), as an alloying element was in the development of (Fe-Si) steels for magnetic sheets. History credits Robert A. Hadfield in England with discovering the magnetic properties of silicon steels around 1900. Fessenden, a researcher at Westinghouse (in the United States), known for his discovery of a rectifier (detector of electromagnetic waves) in 1900, the liquid baretter, had been developing Fe-Si steels in collaboration with ARMCO since 1902 for the transformers and electric motors manufactured by this company, hence, the origin of silicon.

be obtained using materials with a high electrical resistance ("after trial of a large number of substances, pure silicon was found to possess high thermoelectric force against any metal contact, a relatively high and constant resistance and great stability") (Pickard 1906b). In 1906, Pickard set up the Wireless Speciality Apparatus Company (WSA) and sold silicon detectors to all the major radio manufacturers until the 1920s.

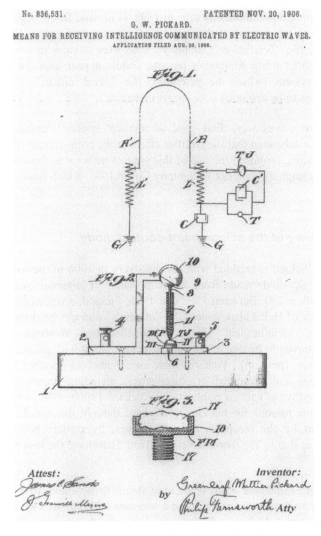

Figure 2.8. *Pickard patent (Pickard 1906a)*

At the same time, in 1906, Dunwoody patented the use of carborundum (SiC) as a rectifier (Dunwoody 1906) to replace galena, while acknowledging that "carborundum appears to be not so good as silicon" (Lee 2009). The American de Forest Wireless Tele-graph company was the first to use carborundum crystal for its telegraph transmissions in 1906. The first transatlantic radio communication by Marconi took place in 1907 using carborundum as a detector.

Point-contact detectors (based on galena) played a major role in the detection of coded messages during the First World War. However, radio-operators had to use trial and error to find the hot spots on the surface of the crystal that would give good reception. It was because of this variability ("such variability bordering on what seemed the mystical, plagued the early history of crystal detectors") (Seitz and Einspruch 1998, p. 28) that they were dethroned by vacuum diodes from the 1920s onwards.

2.2.4. *The germanium point-contact diode*

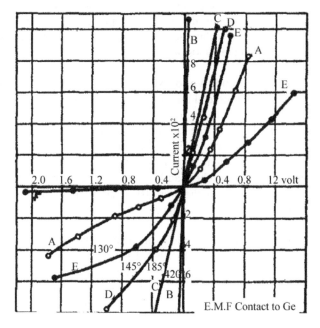

Figure 2.9. *Characteristic curves (current/voltage) of a germanium/iron point-contact diode at different temperatures (Merritt 1925)*

A great deal of work was carried out on these diodes using different materials, including selenium tellurium and germanium, described as a rare metal

(Flowers 1909). Figure 2.9 shows a number of characteristic curves (current/voltage) for germanium point-contact diodes obtained in 1925 by Ernest Merritt, demonstrating the effect of temperature and heat treatment of the material on rectification "quality". For a diode made with the raw germanium available at the time (curve A), the rectifying effect is weak and decreases with temperature (curves E, D, C, B). On the other hand, after heat treatment (heating to 400°C and cooling) (curve F), the diode showed a very clear rectifying effect, and the resistance in the blocked direction was very high (Merritt 1925).

2.2.5. Reception of radar waves

From 1940 onwards, vacuum tubes were in turn superseded by silicon diodes as frequency converters for radar wave reception circuits (frequency converter in microwave reception) (Scaff and Ohl 1947; Torrey and Whitmer 1948, p. 7).

In the years leading up to the Second World War, research was being carried out in various countries (France[4], Germany, the UK and the United States) into ultra-high frequency waves and the means of producing and receiving them. Detecting aircraft using this technique, better known as radar detection, was of interest to the military from the 1930s onwards.

At the start of the Second World War, radar development was sufficiently advanced to be considered of crucial importance in air battles.

The research carried out in the countries mentioned above led to the same solution, namely, the superiority of the silicon/tungsten diode as a frequency converter for signals from radar reception circuits. Researchers, looking for a solution to replace the vacuum tube, had been using galena or silicon diodes either for their own pleasure or during the war, 10 or 20 years earlier. They therefore naturally turned to these diodes as a solution.

In 1936, Hans E. Hollmann, a radio amateur in his youth, was one of the pioneers in the field of microwaves. In 1932–1933, he took part in an international expedition to study the ionized layers of the atmosphere at the North Pole, and he was able to observe the use of microwaves for these studies. Until 1935, he worked at GEMA (Gesellschaft für Elektro-Akustic und Mechanischer Apparat) on the development of a radar system. In 1936, he published a treatise, *Physik und*

---

4 Radar consists of two essential components: the magnetron for transmitting waves and the point-contact diode for receiving reflected waves. France's contribution to radar before the war was essentially in the development of the magnetron: the microwave transmitter.

Technik der ultrakurzen Wellen, in which, in the *Detektorempfang* section, he wrote: "Currently, the component to be used at very high frequencies is the crystal detector; specifically the pyrite crystal (FeS_2) associated with an iron or bronze tip" (Hollmann 1936).

In 1938, Rottgardt (1938), then a researcher at the Institute of Electrophysics of the German Air Force Research Establishment, published the results of an extensive study of different combinations of point-contact diode components, concluding that the silicon/tungsten combination is the most effective for detecting very short wavelengths (50–1.4 cm). "This was clearly one of the landmark studies" (Seitz and Einspruch 1998, p. 97).

After capturing a British bomber in February 1943 and discovering the radar in the aircraft, the German government began research into radar (which it had not believed in until then). A team was set up to develop a silicon detector in the Telefunken laboratories with Herbert Mataré and another team in the aviation laboratories with Heinrich Welker to develop a germanium detector[5]. According to Kai Handel, by the end of the war, German specialists had acquired the same "understanding" of semiconductor physics and their use as their British counterparts (Handel 1998).

In England, the British government, aware of the great threat of German rearmament, and in particular the rapid development of bombers, decided in 1937 that its fighter planes should be equipped with radar. Denis Robinson, who was carrying out television research in a private company, was sent by his government to a secret laboratory, the Telecommunications Research Establishment (TRE), "to develop a detector circuit for reception from a potentially available tencentimeter microwave source". Robinson set about researching the laboratory's library, and it was in Hollmann's book that he found the information about the point-contact diode's ability to detect microwaves. "The only thing we can use as a receiver is the crystal and cat's whisker. [...] Well, I was delighted because that's what I'd used ten years previously", he declared (Seitz and Einspruch 1998, p. 115). Robinson, in collaboration with Skinner, tested a large number of crystal-whisker combinations, concluding that metallurgical-grade silicon combined with a tungsten wire provided the only satisfactory solution. A breakthrough from a practical point of view came on June 16, 1940, when Skinner showed that silicon could be soldered to a tungsten rod, and thus that a silicon-soldered tungsten-tip diode could be encapsulated, the capsule being filled with a viscous liquid to dampen vibrations (Seitz and Einpruch 1998, p. 116). Diodes of this type were subsequently produced by the British

---

5 We will find again Herbert Mataré and Heinrich Welker in the development of the transistor in France after the war (Chapter 3).

Thomson-Houston company and the British General Electric Company. However, the highly variable behavior of these diodes meant that selection had to be made by trial and error. Subsequent British developments are described in section 2.6.

In the years leading up to the Second World War, major work was being carried out at Bell Labs on the use of ultrashort waves for telephone transmissions by ATT (the parent company). According to Riordan, George Southworth was trying to "detect" ultrashort waves (1/10 of a meter) using vacuum tubes, without success, and the same with copper-oxide rectifiers. He then decided to try one of the old cat's whisker crystal detectors found in the radio sets he had used during the First World War, when serving in the Army Signal Corps. He managed to obtain one from a dealer in old radios. After some trial and error, he eventually found hot spots that detected these ultra-short waves. Working with Russell Ohl, who had also been very interested in radio communications as an amateur radio operator since the First World War, he asked Ohl to undertake a search for materials that could be used to make components receive these ultra-short waves. Russell Ohl tested over a hundred materials and confirmed that the silicon/tungsten pairing (which he had used in his college days to make a radio) was by far the most sensitive (Southworth 1936; Riordan and Hoddeson 1997; Seitz and Einspruch 1998, p. 155, p. 162).

From 1941 onwards, point-contact diodes were used during the Second World War, following developments in England and essentially in the United States under the direction of the MIT Radiation Laboratory (Torrey and Whitmer 1948, p. 7).

Developments relating to germanium diodes are described in section 2.3.2.

2.3. Research during the Second World War

2.3.1. *Research on silicon*

2.3.1.1. *Discovery of silicon N and P*

"Ohl made a dramatic discovery".[6]

In 1937, following his work on the choice of the material for the point-contact diode giving the best results for the reception of ultrashort waves, namely silicon, having noted the erratic behavior of the diodes, which he attributed to heterogeneities of composition, therefore to impurities, Russsel Ohl from Bell Labs,

---

6 "An unsung hero of semiconductor history: Russell Ohl, the inventor of the pn junction. Yet the name of its inventor remains largely unknown to the many who benefit from his crucial break-through" (Riordan and Hoddeson 1997, p. 89).

asked a chemist, Grisdale, to prepare samples of "pure silicon". Grisdale, using 99.8% pure silicon powder obtained from the Electrometallurgical Company, prepared small samples by flame fusion. These samples showed much better rectifying characteristics than those obtained with metal metallurgical silicon. Ohl attributed this to the "purity" of the silicon.

In 1939, continuing his research on silicon, he asked metallurgists Jack Scaff and Henry Theuerer to prepare larger samples. Since 1937, Scaff had been involved in studying the role of impurities on the quality of "soft" magnetic materials. The first silicon ingots obtained broke into small pieces during cooling (due to the strong expansion of silicon during solidification). Nevertheless, Ohl found that the rectification characteristics were close to those obtained on the first small samples.

To prevent the ingots from cracking on cooling (severe cracking), silicon ingots were produced by slow solidification in helium (Scaff 1946, 1970). Dense, uncracked ingots were obtained (Figure 2.10). The ingot consisted of three crystal zones: a columnar zone separated into two zones by a thin striated layer (photovoltaic barrier) and an equiaxed central zone. On this ingot, Ohl fortuitously discovered that this columnar zone exhibited the photoelectric effect. This zone was subsequently called the "PN junction", and Ohl's discovery became known as the PN junction (discussed in Chapter 4). Ohl filed three patents in 1941, which were granted in 1946: Ohl, R. (1946a): discovery of silicon N and P silicon; Ohl, R. (1946b): discovery of the PN junction; Ohl, R. (1946c): discovery of the photoelectric effect of a PN junction.

In examining this ingot, Ohl made a "dramatic discovery" according to Scaff[7]. On February 23, 1940 (Scaff 1970), Ohl made a point-contact diode on columnar samples extracted from the outer part of the ingot (the first to solidify). The current I flowed (easy flow) in the direction silicon→ metal tip when the silicon wafer was connected to the + pole of the generator (Figure 2.11(a)), and for columnar samples taken from the central part of the ingots, the current I (easy flow) flowed from the metal tip to the silicon wafer when the silicon wafer was connected to the negative pole of the battery (Figure 2.11(b)). Hence, the names silicon P (p-type), positive silicon, for wafers extracted from the outer part of the ingot and silicon N (n-type), negative silicon, for the pellets extracted from the columnar zone of the central part of the ingot, formulated and reported in a BTL memorandum on January 31, 1941 (Ohl 1946a).

---

7 Scaff (1970) explicitly credits Ohl with the authorship of the discovery: "In examining such an ingot, Ohl made a dramatic discovery".

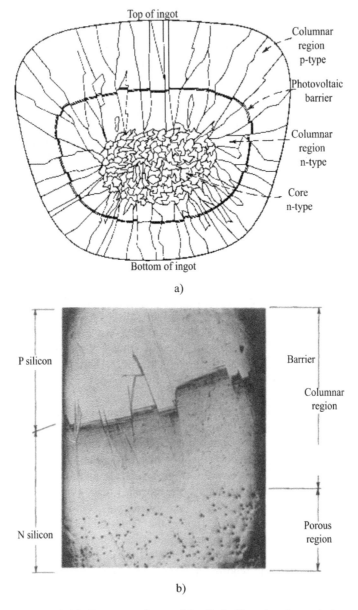

a)

b)

Figure 2.10. *Structure of one of the first silicon ingots made by Jack Scaff (a) cross-section; (b) microstructure at the transition between silicon N and P (Scaff 1970)*

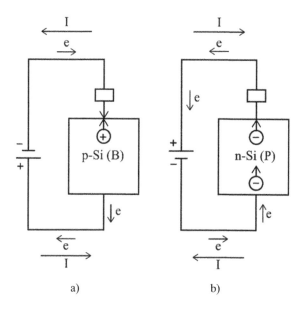

Figure 2.11. *Majority carrier currents: (a) holes in a silicon P-based (p-Si) diode; (b) electrons in a silicon N-based (n-Si) diode*

In a silicon P diode, the majority current is a current of holes.

In a silicon N diode, the majority current is an electron current (see analysis in the Appendix, section 2.5).

The characteristic curves obtained by Ohl are shown in Figure 2.12.

Ohl, Scaff and Theuerer gradually became convinced that these behaviors were due to the segregation of impurities along the ingot during (slow) solidification. "A number of things suggested to us that these unusual effects in the silicon ingot were related to impurities".

Through slow solidification, the impurities segregate along the length of the ingot, so that the fraction of the ingot that solidifies last contains the impurities.

The Si "columnar region p-type" samples in the outer zone of the ingot, the first to solidify and therefore the purest, were rich in boron, which practically does not segregate ($k = 0.8$).

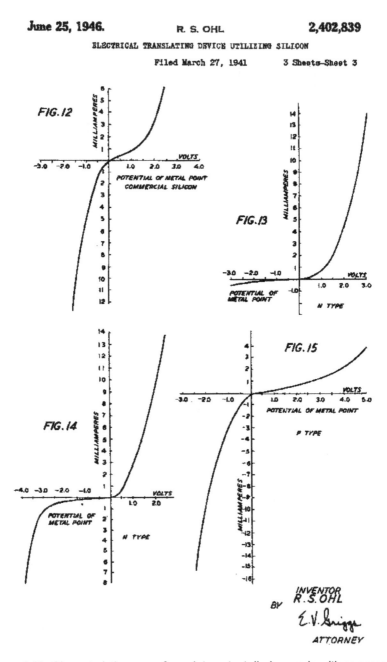

June 25, 1946.   R. S. OHL   2,402,839

ELECTRICAL TRANSLATING DEVICE UTILIZING SILICON

Filed March 27, 1941   3 Sheets—Sheet 3

Figure 2.12. *Characteristic curves for point-contact diodes made with commercial silicon (Figure 12), silicon N (Figures 13 and 14), and silicon P (Figure 15) extracted from the ingot (Figure 2.10(a) of this work) (Ohl 1946a)*

Si "columnar region n-type" samples on the other side of the photovoltaic barrier (Figure 2.10(a)) had a phosgene gas smell, which was attributed to the presence of phosphorus. Phosphorus segregates strongly in the liquid (k_{eff} = 0.35), so it is concentrated in the last region to solidify. The addition of phosphorus to molten silicon confirmed that phosphorus was the impurity responsible for the n-behavior of phosphorus-doped silicon. As reported by Scaff, Ohl and Scaff then remembered that a silicon sample showing the same effect as a p-type sample had been melted in an alumina crucible, and concluded that its behavior must be due to the presence of aluminum.

For this reason, in Chapter 1, silicon N is referred to as N (n-doped) silicon and silicon P as P (p-doped) silicon.

These observations led to the study of the effect of the addition of numerous group III and V elements on the electrical behavior of silicon, in the laboratory of Professor Frederick Seitz at the University of Pennsylvania; then in 1942, in the laboratory of Professor Lark-Horovitz at Purdue University on the behavior of germanium.

These observations about the role of specific impurities (such as boron and phosphorus) on the conductivity of silicon were indeed the "discovery" that enabled the research undertaken following this discovery to establish in 1942 that germanium and silicon were semiconducting materials whose conductivity could be controlled by doping. It was Lark-Horovitz of Purdue University, through his in-depth studies of germanium, to whom history attributes the definitive characterization of these two elements as semiconductors (see section 2.3.2). Until then, materials classified as semiconductors, because of the variation of their conductivity with temperature opposite to that of metals, were compounds: sulfides (galena PbS) or oxides (Cu_2O, ZnO).

Variations in the conductivity of different oxide samples at the same temperature were attributed to deviations in composition from the stoichiometry. For example, in the case of Cu_2O, conductivity increased with increasing oxygen content (obtained on samples subjected to increasing oxygen pressures), while in the case of ZnO, conductivity increases with decreasing oxygen content. These variations in conductivity are due to an excess of positive or negative ions in relation to stoichiometry. In the case of copper oxide, the current is carried by holes; in the case of zinc oxide, the current is carried by electrons. It should be noted here that in the case of PbS (galena), the material used in the diodes housed within radio receivers, the current can be carried by electrons or holes, depending on the deviation from stoichiometry (Busch 1989).

The basic property of semiconductors – the ability to modify and control electrical conductivity by adding "impurities" – was discovered.

2.3.1.2. Production of 5N purity silicon

Only metallurgical silicon (Chapter 1) was available in 1939, and the point-contact diodes manufactured with this silicon did not offer very high performance, as seen from the curves shown in Figure 2.13: small difference between the characteristic curves in the forward direction (F curve) and in the reverse direction (R curve).

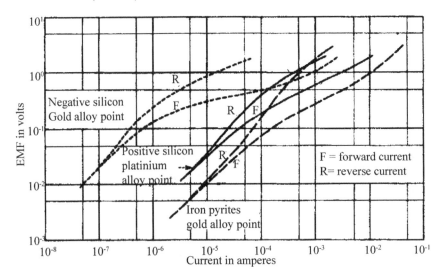

Figure 2.13. *Characteristics of N (negative silicon) and P (positive silicon) diodes taken from the ingot produced in 1940 (Figure 2.10) and of iron pyrite diodes (Scaff and Ohl 1947, p. 7)*

A silicon purification process (silicon 5N (99.999) (0.001%)) was then developed by Olson (1988) in the laboratories of the DuPont de Nemours company in Wilmington, Delaware (E. I. du Pont de Nemours, who founded the company in 1802, was a pupil of Lavoisier). Olson was involved in a program to develop pigments to replace titanium oxide pigments (TiO_2)[8]. In search of a new silica-based

---

8 At the same time, the industrial process for reducing titanium chloride by magnesium was developed in 1939 by W. J. Kroll in the laboratories of Union Carbide in Niagara Falls (New York). Reduction by sodium had been discovered in 1910 by M. H. Hunter (Hunter process). Union Carbide was involved in a program to develop the production of ultrapure silicon for the solar cell program (Volume 2, Chapter 4).

pigment, he was working on obtaining this pigment from purified silicon obtained by reacting gaseous silicon chloride $SiCl_4$[9] , purified by distillation, with gaseous zinc (purified 5N):

$$SiCl_4 (g) + 2Zn(g) \rightarrow 2ZnCl_2 (g) + Si(s) \text{ (crystallized particles)}$$

Zinc silicide is not produced, due to the low chemical affinity between silicon and zinc, and so a high-purity product is obtained. The silicon obtained by this gas-phase reaction was a mass of needle-like crystals with a purity of 99.999 (less than 0.001% impurities).

In 1855, Henri Sainte-Claire Deville had achieved the direct preparation of crystalline silicon "which is the analogue of diamond" by reducing silicon chloride with sodium (see Chapter 1, section 1.3).

Frederick Seitz, then a consultant to Du Pont de Nemours on these pigment problems and at the same time acting as an adviser to MIT's Radiation Laboratory, immediately realized the value of this silicon for the manufacture of diodes for radars, and informed the Radiation Laboratory (Seitz and Einspruch 1998, p. 129).

As a result, in early 1942, DuPont de Nemours became the supplier of 5N (99.999) purity silicon throughout the war to companies such as ATT, Westinghouse, Sylvania and General Electric for the manufacture of diodes for radars.

It turns out that a "defect" in the DuPont de Nemours process ($SiCl_4$ + Zn) process – that of not eliminating the boron – proved beneficial for the sensitivity of the radar point-contact diode. The boron chloride (BCl_3) present was reduced to boron, simultaneously with the $SiCl_4$, by zinc and incorporated into the silicon produced.

Henry Theuerer of Bell Labs showed that the addition of boron (of the order of 0.0015% to 0.005% by weight) to this silicon produced silicon-based P diodes with high sensitivity (Theuerer 1943; Scaff and Theuerer 1949).

However, although a considerable amount of exploratory testing was carried out, it was not possible to determine why some dopants were better than others in terms of the characteristics of a "mixer" diode.

---

9 Silicon chloride was known to chemists as a volatile liquid used during the First World War as an agent to produce smoke screens.

After the war, to meet the purity requirements needed to develop transistors, purities in the ppb range were required. DuPont developed a production line based on the decomposition of silane on a silicon rod heated to incandescence. DuPont was a supplier of high-purity silicon until 1961 (section 2.2, Chapter 4).

2.3.1.3. *Silicon oxidation treatment: the "gettering effect"*

An oxidation treatment of the silicon wafer surface, followed by removal of the oxide layer by hydrofluoric acid attack, initially developed in England by the General Electric Company, was applied in the United States after being perfected (steam oxidation) by Jack Scaff in 1944.

The oxidation treatment of silicon proved to be the most effective way of improving the reception of radar waves by the point-contact diode.

The current-voltage characteristic curves (Figures 8 and 9 in Figure 2.14(b)) show that this treatment makes it possible to obtain reverse resistances (in the blocked direction) several thousand times greater than the direct resistances and a very high breakdown voltage (Figure 9 in Figure 2.14(b)).

The effectiveness of this treatment is due to the purification of a surface layer of silicon by extraction of impurities (including dopant) (the impurities diffuse into the oxide layer during heat treatment), thus producing a surface layer of very high resistance. This purification by extraction of impurities is called the "gettering effect".

> No innovation in crystal manufacturing technique has contributed so much to improve crystal performance as has the heat-treatment process. (Torrey and Whitmer 1948, p. 10)

NOTE.– The beneficial role of the oxidation treatment of silicon in the oxidation masking process (see Chapter 5, section 5.3.1), which led to the development of the bipolar transistor with a planar structure (see Chapter 5, section 5.3.3), an achievement which led to the development of integrated circuits based on these transistors.

The oxide layer plays an even more crucial role as a component of the MOSFET transistor (Chapter 6, section 6.1.5), in particular with regard to the "gettering effect" produced by an oxide surface layer in formation and on the purification of the surface layer of solar cells by this effect (Volume 2, Chapter 4, section 4.5.2).

10
1
0.1
0.01
VOLTS
0.0001
0.001
0.01
0.1
1
10
100
MILLIAMPERES
REVERSE
FORWARD
BEFORE OXIDIZING
AFTER OXIDIZING

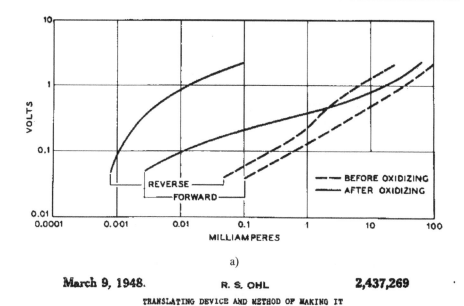

a)

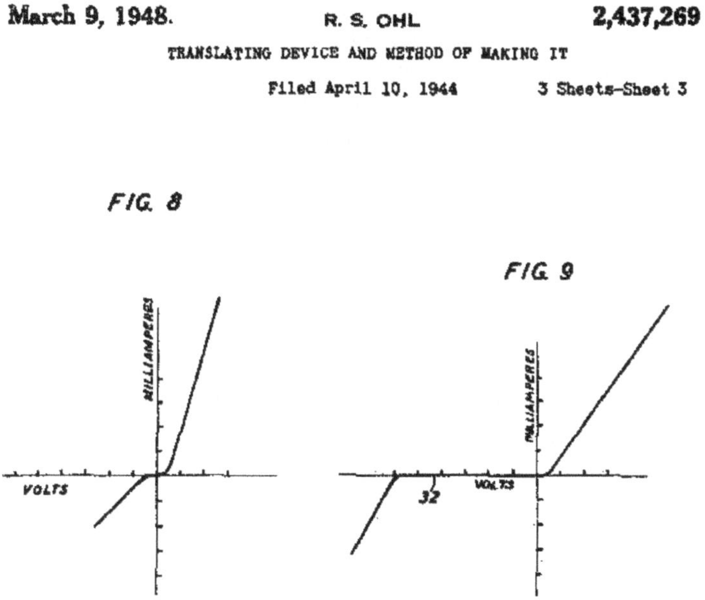

March 9, 1948.
R. S. OHL
2,437,269
TRANSLATING DEVICE AND METHOD OF MAKING IT
Filed April 10, 1944
3 Sheets-Sheet 3
FIG. 8
MILLIAMPERES
VOLTS
FIG. 9
MILLIAMPERES
VOLTS
32

b)

Figure 2.14. *Effect of oxidation treatment on the characteristic of a point-contact diode: (a) in P silicon doped with 0.003% boron (Scaff 1970); (b) Figure 8 before TT and Figure 9 after TT from Ohl"s patent (Ohl 1946a)*

2.3.2. *Research on germanium*

Intensive research on the properties of germanium by physicists at Purdue University and in the laboratories of General Electric by Harper Q. North was carried out during the Second World War. This research was motivated by the fact that if silicon behaved so well as a rectifier, it was essential to study the element located in the same column of the Mendeleev table. The research group led by Karl Lark-Horovitz of Purdue University had obtained a contract from the MIT Radiation Laboratory (coordinator of all research into "radar") in March 1942 to study galena as a rectifier.

According to his writings, Gordon Teal from Bell Labs, recruited in 1930, was the initiator of research on germanium from February 1942 onward and produced the first germanium-based component consisting of a deposit of germanium on a tantalum filament, by thermal decomposition of the gas "digermane" Ge H^{26}, obtained from Brown University where he had completed his thesis. He was authorized to carry out a research program on germanium-based diodes and presented Professor Lark-Horovitz with the characteristics of his diodes. Lark-Horovitz was "impressed" and decided to concentrate his research on germanium rather than galena (Teal 1976). According to Benzer (1949), germanium-based rectifiers were found to be useful in microwave frequencies in 1942 by the Sperry Research laboratories, which induced the Purdue University research group to study germanium.

2.3.2.1. *Germanium production*

At the suggestion of Lark-Horovitz, the company Eagle-Picher developed a process for producing germanium oxide GeO_2 as a by-product of its tin production. The same company also supplied $GeCl_4$.

From the oxide, the Lark-Horovitz team, General Electric and Bell Labs developed a hydrogen reduction process. Using dry hydrogen reduction at 650°C for 3 h, Bell Labs obtained germanium with conductivities ranging from 4 to 12 Ω·cm. Using wet hydrogen reduction, General Electric obtained germanium with a resistivity of 40 Ω·cm (the higher the purity, the higher the resistivity).

From germanium chloride purified by distillation, DuPont de Nemours produced germanium by reduction with gaseous zinc (using the same process as for silicon) at 930°C (presented above). Treatment with hydrochloric acid separated the residual zinc from the germanium. Hydrofluoric acid etching removed the residual silica. Using this method, the germanium obtained had an impurity content of 0.05% and a zinc content of 0.2%, and a resistivity of around 30–40 Ω·cm. Theuerer and Scaff of Bell Labs later showed that, by controlled solidification of germanium obtained by the DuPont de Nemours process, the purified zones at the top of the ingot showed resistances of 120 Ω·cm (Torrey and Whitmer 1948, p. 304).

One of the main outcomes of this extensive germanium research program was the production of germanium of such purity and (coarse) crystalline structure, which enabled Bardeen and Brattain to discover the point-contact transistor (Chapter 3).

2.3.2.2. Germanium point-contact diodes: characteristic curves

Purdue University doped purified germanium with B, Al, P, As, Sn and Sb and studied their influence on rectification properties. This work led to the same conclusions as for silicon: the elements B, Al, Ga and In produced a conductivity of the p type and the elements As, Sn and Sb produced conductivity of the n type.

While silicon N (n-type) (n-doped) and silicon P (p-type) (p-doped) exhibited the rectifier effect, only diodes made with germanium N (n-type) (n-doped) with the three elements (As, Sn, Sb) showed the rectifying effect. With germanium P(p-type) (p-doped), the contact is ohmic. As shown in section 2.1.2, these different behaviors are due to different energy bandgaps (E_G = 1.12 eV for silicon and E_G = 0.67 eV for germanium).

In addition, Theuerer and Scaff (1951) showed that tin-doped germanium, as solidified, was n-type and, after heat treatment at 800°C followed by rapid cooling, p-type, and that the effect was reversible.

Figure 2.15 shows the characteristic curves of two point-contact diodes based on germanium N and P (both doped with tin (0.1%)). The N diode is a rectifier. Note that the breakdown voltage is very high. Diode P has no rectifier effect, and the contact is ohmic.

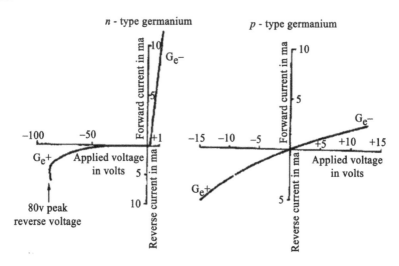

Figure 2.15. *Typical characteristic curves for germanium-based diodes with tungsten tips (Scaff and Theuerer 1945)*

The influence of germanium purity and heat treatment on breakdown voltage of tin (n)-doped germanium diodes was studied by Bell Labs, who definitively established the effect of germanium purity on breakdown voltage. The higher the purity of the germanium, the higher the breakdown voltage. Breakdown voltages of up to 250 V were achieved, enabling them to withstand high-voltage surges while offering low resistance under positive polarity.

2.3.2.3. Current in a germanium-based diode N

In a silicon N diode, the current flowing in the forward direction is mainly a current of electrons (Figure 2.11(b)); in the germanium N diode, the current is made up of an electron current and a hole current. This mechanism was postulated and experimentally revealed by Bardeen and Brattain (1949). The expressions for the currents in an N germanium diode are given in Appendix 2.5.

If the current in a germanium N diode had been predominantly a current of electrons, as in a silicon N diode, the point-contact transistor would not have existed and therefore would not have been discovered (Chapter 3, section 3.2.2).

2.3.2.4. Point-contact diode produced by welding the electrode to the germanium N wafer

From 1942, Harper Q. North in the laboratories of General Electric continued research into the N antimony-doped germanium diode, as part of the MIT-led program. Significant progress was made in the quality (high breakdown voltage) and reliability of these diodes. Reliability of these germanium N diodes was achieved by welding the Pt-10%Ru metal tip to the germanium pellet (North 1946).

The improved properties of these germanium N-based diodes due to this soldering is due to the formation of a P-zone in the area of contact with the metal tip by the heat treatment carried out by the solder in this area, in accordance with what Theuerer and Scaff had shown (see section 2.3.2.2). In fact, the solder produced a PN diode (Van Vasseur 1955) (see Chapter 4).

These germanium N diodes proved to be excellent rectifiers, hence their widespread use after the war. On the other hand, their characteristics were inferior to those of silicon diodes for use as mixers; that is, for receiving radar microwaves.

These germanium N diodes, with a high breakdown voltage due to the high purity of germanium, were used as components of logic circuits in computers built from the 1950s onward because of their very high switching speed (see section 2.4.1).

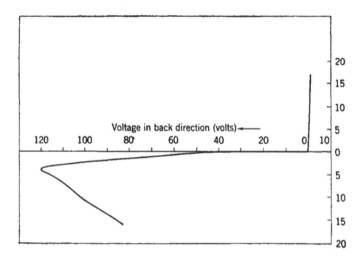

Figure 2.16. *Characteristic curve (current-voltage) of an N germanium-based diode with high breakdown voltage (Torrey and Whitmer 1948, p. 23)*

2.4. The industrial development of germanium diodes after the Second World War

According to Torrey and Whitmer, following the demonstration of the high breakdown voltage property, these diodes were the subject of tremendous interest because of their possible use in various applications at frequencies of 30 MHz or lower in the circuits of radio sets and television sets (Torrey and Whitmer 1948, p. 10).

During the war, many companies (General Electric, Sylvania and Raytheon) had researched and manufactured such diodes. Their characteristics were well known to them, so industrial development naturally followed.

According to Seitz and Einspruch, Harper North from the General Electric laboratory, who had developed the whisker solder, "decided that germanium was a better material than silicon". They also mentioned that "Germanium was the obvious choice. [...] It was available in large quantities, as good sources had been found during the war. It could be purified to a high degree, and processes for obtaining single crystals were available. Germanium diodes were highly reliable and mechanically stable, achieved by soldering the tip to the germanium pellet" (Seitz and Einspruch 1998, pp. 130 and 198).

Commercial production of germanium N (doped with Sn tin) began in 1946 as a replacement for vacuum diodes. Sylvania was the first company to start industrial production of the IN34 diode in 1946. The range of diodes produced by Sylvania is shown in Figure 2.17. A number of companies (vacuum diode manufacturers), including General Electric, Raytheon, Philips and Telefunken, began producing germanium point-contact diodes (Ward 2008). Ranges of diodes with specific properties were marketed by these companies[10]. The first diode produced by General Electric was marketed in 1948. These diodes were of the whisker (Pt-10%Ru) welded/germanium doped with antimony (Sb) type. Philips began manufacturing diodes with an arsenic-doped N germanium tip and a W tip welded by applying pulses of high current. Figure 2.18 shows the characteristic curves of various diodes manufactured by Philips, each with a specific application.

Alongside these major companies, others were set up in the 1950s. Transitron was founded in 1952 by David Balakar, a Bell Labs defector, who developed a gold-bonded germanium diode (Balakar 1958), "a breakthrough semiconductor device" (Ward 2010). Transitron was one of the two or three largest producers of such diodes until 1965.

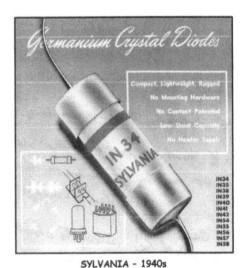

SYLVANIA - 1940s

Figure 2.17. *Germanium point-contact diodes produced by Sylvania (Ward 2014). For a color version of this figure, see www.iste.co.uk/vignes/silicon1.zip*

---

10 Sylvania published a brochure (for a fee) entitled "40 Uses for Germanium diodes". Philips published an article in its *Philips Tech Review* in February 1955 entitled "The application of point-contact germanium diodes", by J. Jaeger.

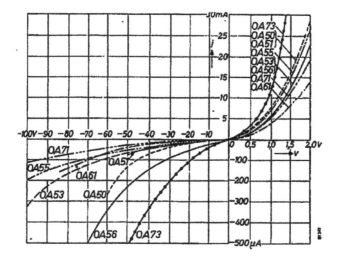

Figure 2.18. *Characteristic curves for various Philips germanium diodes (Jaeger 1955)*

At the end of the 1940s, the market for these diodes was growing by the hundreds of thousands every year, following the development of television and portable radio, whose circuits required this type of component. GE was manufacturing around 100,000 diodes a day at the end of the 1950s (Sheckler 2004). The manufacture of these diodes continued throughout the 1950s and beyond.

In the United States, 16.5 million germanium diodes were sold in 1954. In 1959, sales of germanium diodes reached 65 million units, and in 1965, 383 million units. The majority of these diodes were gold-bonded switching diodes (Seidenberg 2014).

In addition, the specific characteristics of these diodes (very high switching speed) compared with vacuum diodes and bipolar transistors made it possible to produce the first computers with logic circuits made up of germanium point-contact diodes.

The first computers whose logic functions were implemented by circuits based on germanium diodes (all diode logics) appeared: MADDIDA in 1949, SEAC in 1950 and IBM 1401 in 1960 (Reed 2006).

The MADDIDA (Magnetic Drum Digital Differential Analyzer) computer, built in 1949 by the Northrop Aircraft Corporation, had 53 vacuum tubes and a matrix of 904 germanium-based diodes (Dorian 1950).

The SEAC (National Bureau of Standard Eastern Automatic Computer) in 1950 had 747 vacuum tubes and 10,500 germanium diodes. "All computing and switching are performed by interconnected germanium crystal diodes rather than by electron tubes" (SEAC 1950).

In 1954, TRADIC, the first transistorized military computer, built by Bell Labs for the US Air Force, had 684 Ge point-contact transistors and 10,358 Ge point-contact diodes, a third of which were used as logic circuit components (Irvine 2001).

The central processing unit of the IBM 1401 computer, released in 1960, had 10,600 germanium bipolar transistors (alloy junction) (Chapter 5) and 13,200 germanium point-contact diodes.

2.5. Appendix: currents in a metal–semiconductor diode

For a metal-semiconductor diode N, exhibiting the rectifier effect, $\Phi_m > \Phi_{s(n)}$ (Figure 2.2(b)), for a positive and increasing potential V_F applied to the metal electrode (Figure 2.11(b)), the potential barrier ($V_{bi} - V_F$) decreases and the thickness of the depletion layer decreases, and the current I in the diode (conducting direction) increases (Figure 2.1). The depletion layer and the potential barrier disappear at an applied potential $V = V_{bi}$. Above this, the current I increases linearly with the applied voltage, the circuit resistance being that of a silicon wafer.

For a negative potential applied to the metal electrode, the potential barrier and thickness of the barrier layer increase; the resistance of this "barrier layer" R becomes very high, and so the current through the diode decreases rapidly and becomes very low.

In a diode (metal-semiconductor N or P (n or p doped)), the current that flows is the sum of a current of electrons (conduction) current and a current of holes flowing in opposite directions.

In a metal-N semiconductor diode, the current of electrons flowing from the semiconductor to the metal electrode (Figure 2.11(b)) is due to thermoelectronic emission, in both the forward and reverse directions. This is expressed by[11]:

$$J_n = J_s.\{\exp(eV/kT) - 1\} \text{ with: } J_S = AT^2.\exp(-e\,\Phi_{Bn}/kT) \tag{2.2}$$

---

11 All the expressions (formulae) quoted are taken from Sze (2002, p. 225) and explained in Volume 1, Chapter 4 (Appendix given in section 4.3).

where J_S is the "saturation current" (for a potential of 0.2 V, the current $J_n = 0.1$ A/cm^2) and Φ_{Bn} is the potential barrier to the passage of electrons from the semiconductor to the metal (Figure 2.2(b)).

For the hole current in the absence of an applied potential, at the limit of the depletion zone (layer) in the N semiconductor, the concentrations of the charge carriers satisfy the mass action law (formulae [1.5] and [1.6], Chapter 1), which fixes the concentration of minority charge carriers at this limit:

$$p_{(N)0}.n_{(N)0} = p_{(N)0}.N_D = n^2_i \qquad\qquad [2.3]$$

If a positive potential is applied to the metal electrode, the potential barrier $(V_{bi} - V_F)$ decreases and the thickness of the depletion layer decreases; the concentration of minority carriers (the holes) at the limit of the depletion zone becomes:

$$p_{(N)} = \{n_i^2/N_D\} \, \exp\,(eV_F/kT) > p_{(N)0} \qquad\qquad [2.4]$$

The excess concentration of these carriers $p_{(N)}$ over the equilibrium concentration $p_{(N)0}$ is interpreted as an "injection" of "holes" (minority carriers) by the metal electrode.

In fact, electron-hole pairs are generated in the semiconductor: a valence electron passes into the conduction band, is extracted by the electrode, and a hole remains at the limit of the depletion zone (Chapter 1, Figure 1.10(a)).

The difference in concentration $(p_{(N)} - p_{(N)0})$ between the boundary of the depletion zone and the core of the wafer induces a diffusion flux of holes, given by the first "law of diffusion" (Fick's law), and therefore a current of holes inverse of the electron current (Chapter 4, formula [4.6]).

As these holes move away from the metal electrode, they gradually recombine with the electrons (Chapter 1, section 1.2.6). The resulting hole current has the expression (this formula is established in Chapter 4, section 4.3.2):

$$J_p = J_{p0} \, \{\exp\,(eV/kT) - 1\} \text{ with: } J_{p0} = e \, D_p n_i^2/(L_p.N_D) \qquad\qquad [2.5]$$

This hole current is proportional to the square of the concentration of intrinsic carriers: n_i^2, that is, to $\exp(-E_G/kT)$ (Table 1.1 and expression [1.2] in Chapter 1). For silicon, $n_i^2 = 10^{20}$, and for germanium, $n_i^2 = 4 \times 10^{26}$.

As a result, for a silicon N diode, this hole current is several orders of magnitude smaller than the electron current. The ratio between the currents J_p/J_n is of the order of 10^{-7}. The current flowing in such a diode is practically an electron current (see Figure 2.11(b)).

On the other hand, for a N germanium-metal diode which has a small bandgap $E_G = 0.67$ eV, this ratio is much higher. The current is the sum of an electron current and a current of holes.

The discovery of the point-contact transistor (Chapter 3) made on germanium N wafer is due to this characteristic of a germanium N-metal diode: a current is made up of a current of electrons and a current of holes, itself a consequence of the low bandgap of the germanium $E_G = 0.72$ V.

On the other hand, such a silicon-based point-contact component would not have had the amplifying effect of a transistor (see Chapter 3).

2.6. References

Balakar, D. (1958). Crystal diode. Patent, US2832016.

Bardeen, W.H. and Brattain, J. (1949). Physical principles involved in transistor action. *Physical Review*, 75(8), 1208–1220.

Benzer, S. (1949). High inverse voltage germanium rectifiers. *Journal of Applied Physics*, 20, 804–810.

Bethe, H. (1942). Theory of high frequency rectification by silicon crystals. Report, MIT Radiation Laboratory, R.L-184.

Bose, J.C. (1904). Detector for electrical disturbances. Cat's whisker type. Patent, US755840.

Brattain, W.H. and Bardeen, J. (1948). Nature of the forward current in germanium point contacts. *Physical Review*, 74(2), 231–232.

Braun, K.F. (1874). Ueber die stromleitung durch schwefelmetalle. On current conduction through metal sulfides. *Annalen der Physik und Chemie*, 153, 556–563.

Braun, K.F. (1909). Electrical oscillations and wireless telegraphy. Nobel Lecture, 11 December [Online]. Available at: https://www.nobelprize.org/prizes/physics/1909/braun/lecture/.

Busch, G. (1989). Early history of the physics and chemistry of semiconductors – from doubts to fact in a hundred years. *European Journal of Physics*, 10(4), 254–264.

Dorian, J. (1950). Maddida. Report, GM.545.

Dunwoody, H.H.C. (1906). Wireless telegraph system. Patent, US857616.

Emerson, D.T. (1998). The work of J.C. Bose, 100 years of mm-wave research [Online]. Available at: https://ieeeplore.ieee.org.

Flowers, A.E. (1909). Crystal and solid contact rectifiers. *Physical Review*, 29(5), 445.

Georges, F. and Mauviard, F. (1997). Ferdinand Braun, Itinéraire d'un Nobel Cathodique [Online]. Available at: www.cathodique.net.

Handel, K. (1998). The uses and limits of theory: From radar research to the invention of the transistor. Annual Meeting of the History of Science Society at Kansas City (Missouri) [Online]. Available at: https://sites.google.com/site/transistorhistory/Home/european-semiconductor-manufacturers/history-of-transistors-in-france/the-transistron.

Hellgaard, H. (2023). Cat's whisker detector [Online]. Available at: https://commons.wikimedia.org/wiki/File:Kristallradio_(3).jpg.

Hollmann, H.E. (1936). *Physik und Technik der ultrakurzen Wellen*. Springer, Berlin.

Irvine, M.M. (2001). Early digital computers at Bell Telephone laboratories. *IEEE Annals of the History of Computers*, 23(3), 22–42.

Jaeger, J. (1955). The application of point-contact germanium diodes. *Philips Technical Review*, 16(8), 225.

Kuphaldt, T.R. (2009). *Lessons in Electric Circuits. Volume III: Semiconductors*. 5th edition. [Online]. Available at: https://www.academia.edu.

Lee, B. (2009). How Dunwoody's chunk of "Coal" saved both de Forest and Marconi. *AWA Review*, 22(1), 135–146.

Merritt, E. (1925). On contact rectification by metallic germanium. *Proceedings NAS*, 11(12), 743–749.

North, H.Q. (1946). Properties of welded contact germanium rectifiers. *J. Applied Physics*, 17, 912.

Ohl, R. (1946a). Electrical translating device utilizing silicon. Patent, US2402839.

Ohl, R. (1946b). Alternating current rectifier. Patent, US2402661.

Ohl, R. (1946c). Light-sensitive electric device. Patent, US2402662.

Ohl, R. (1948). Translating device and method of making it. Patent, US2437269.

Olson, C.M. (1988). The pure stuff. *American Heritage of Invention and Technology*, 4(1), 58–63.

Orton, J. (2009). *The Story of Semiconductors*. Oxford University Press, Oxford.

Pickard, G.W. (1906a). Means for receiving intelligence communicated by electric waves. Patent, US836531.

Pickard, G.W. (1906b). Thermoelectric wave detectors. *Electrical World*, 11, 1003.

Pickard, G.W (1909). Solid rectifiers. *Electrical Review and Western Electrician*, 54, 343.

Pickard, G.W. (1919). How I invented the crystal detector. *Electrical Experimenter*, 8, 325.

Reed, J.S. (2006). The dawn of the computer age. *Engineering and Science*, 69(1), 7–12.

Riordan, M. and Hoddeson, L. (1997). *Crystal Fire: The Invention of the Transistor and the Birth of the Information Age.* W.W. Norton & Company, New York.

Rottgardt, J. (1938). Flame temperatures vary with knock and combustion-chamber position. *Zeitschrift für Technische Physik*, 19, 262.

Scaff, J.H. (1946). Preparation of silicon materials. Patent, US2402582.

Scaff, J.H. (1970). The role of metallurgy in the technology of electronic materials. *Metallurgical Transactions*, 1(3), 561–573.

Scaff, J.H. and Ohl, R.S. (1947). Development of silicon crystal rectifiers for microwave radar receivers. *The Bell System Technical Journal*, 26(1), 1–30.

Scaff, J.H. and Theuerer, H.C. (1945). Final report on preparation of high back voltage germanium rectifier. Report, NDRC 14-555 BTL.

Scaff, J.H. and Theuerer, H.C. (1949). Translating material of silicon base. Patent, US2485069.

Scaff, J.H., Theuerer, H., Schumacher, E.E. (1949). P-type and N-type silicon and the formation of the photovoltaic barrier in silicon Ingots. *Metals Transactions*, 185, 383.

SEAC (1950). Standard Eastern Automatic Computer. National Bureau of Standards. *Computer Development at the NBS Circular 551 – Technical News Bulletin*, 34(9).

Seidenberg, P. (2014). Archives: From germanium to silicon. A history of change in the technology of the semiconductors [Online]. Available at: https://ethw.org.

Seitz, F. and Einspruch, N.G. (1998). *Electronic Genie. The Tangled History of Silicon.* University of Illinois Press, Illinois.

Sharpless, W.M. (1959). High frequency gallium arsenide point-contact rectifiers. *The Bell System Technical Journal*, 1, 259.

Sheckler, A.D. (2004). The beginning of the semiconductor world. General Electric's role. In *IEEE Conference on the History of Electronics*, IEEE History Center at Rutgers University.

Southworth, G.C. (1936). Hyper-frequency wave guides – General considerations and experimental results. *Bell System Technical Journal*, 15, 284–309.

Sze, S.M. (2002). *Semiconductor Devices*, 2nd edition. Wiley, New York.

Teal, G.K. (1976). Single crystals of germanium and silicon – Basic to the transistor and integrated circuit. *IEEE Transactions on Electron Devices*, 23(7), 621–639.

Theuerer, H.C. (1943). Preparation and rectification characteristics of boron-silicon alloys. BTL Report, 3MM-43-120-75.

Theuerer, H.C. and Scaff, J.H. (1951). Transactions AIME. *Journal of Metals*, 1, 189.

Torrey, H.C. and Whitmer, C.A. (1948). *Crystal Rectifiers.* McGraw Hill, New York.

Van Vasseur, J.C. (1955). The theory and construction of germanium diodes. *Philips Technical Review*, 16(8), 213–224.

Ward, J. (2008). History of crystal diodes. Volume 1: 1950s germanium radio detectors. Transistor Museum [Online]. Available at: www.semiconductormuseum/museumlibrary.

Ward, J. (2010). David Balakar. Historic profile, Transistor Museum [Online]. Available at: www.semiconductormuseum/museumlibrary.

Ward, J. (2014). Diodes à pointe, Sylvania. Transistor Museum [Online]. Available at: http://www.transistormuseum.com.

The Point-Contact Transistor

Figure 3.1. *Replica-of-first-transistor (http://www.circuitstoday.co).*
For a color version of this figure, see www.iste.co.uk/vignes/silicon1.zip

The discovery that silicon conductivity could be controlled by doping (Chapter 2) led, at the end of the Second World War, to research on the effect of an electric field on this conductivity. Although a direct field effect (modulation (increase) of the conductivity of a thin surface layer of a block of germanium N produced by a field of positive polarity) had been demonstrated, this research would only later lead to the realization of the MOSFET field-effect transistor (Chapter 6).

However, this research led to the discovery of the amplifying effect presented by the germanium N point-contact transistor. Such a silicon-based component would not have exhibited the amplifying effect.

The mechanism demonstrated for the operation of this transistor led to the invention of the bipolar transistor by Shockley (Chapter 5).

All the tests leading up to this discovery were carried out with high-purity, coarse-crystalline germanium N.

From 1953 to 1960, prototype computers were built using germanium-based N.

This chapter presents:

– the field effect;

– the ongoing research into the field effect;

– the discovery of the germanium N point-contact transistor;

– the industrial development of the germanium N point-contact transistor.

3.1. The field effect

3.1.1. *"Direct" field effect, "inverse" field effect*

The field effect consists of "polarization": modulation or inversion of the conductivity of a thin surface layer of a doped semiconductor wafer, induced by the application of an electrostatic field perpendicular to the surface of the semiconductor wafer, which is via a flat electrode close to and parallel to the surface of the wafer; the electrode being separated from the semiconductor wafer by a vacuum or a layer of dielectric (Figure 3.2).

For an N semiconductor and a positive polarity at the electrode, a surface layer is enriched in electrons (majority carriers) (Figure 3.2(b)). If a current is passed through this film, the conductivity in the film is increased, and the current flow is increased (majority carrier current). The result is an amplifier effect. This is the "direct field effect" (majority carrier modulation).

For an N semiconductor and a negative polarity at the electrode, the electrons in the surface layer of the N semiconductor are repelled (Figure 3.2(c)), leaving a layer of positive charges (the donor ions). This "depletion layer" increases in thickness with the applied voltage. Above a certain negative voltage, a hole-enriched layer is formed (Chapter 6, Figure 6.2) of very low thickness, called the "inversion layer".

This is the "inverse field effect" (minority carrier modulation). The inversion layer can be formed by "thermal generation of electron-hole pairs" and charge carrier separation (Chapter 1, section 1.2.5), or in the presence of light energy, by the photoelectric effect (Volume 2, Chapter 4), which generates electron-hole pairs in the semiconductor; the polarization-dependent electric field attracts either the holes or the electrons thus generated to the wafer surface.

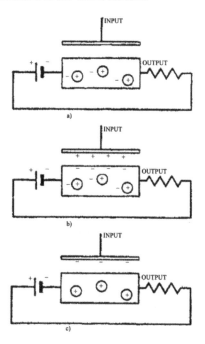

Figure 3.2. *The field effect as displayed in Shockley's (1976, p. 605) article*

The direct field effect (majority carrier modulation) was observed by Bell Labs researchers as early as 1947, but the studies did not lead to the realization of a "transistor", as a result of the "surface states" problem described below (section 3.1.2.1). On the other hand, the series of tests carried out to demonstrate this field effect led to the realization of a transistor operating on a totally different principle: the point-contact transistor (section 3.2).

Field-effect studies led to two types of field-effect transistor: thin film transistor (TFT) and MOSFET.

It was not until 1960 that a transistor based on this direct field effect (the TFT) was developed; and in 1979, a TFT transistor based on amorphous silicon was produced, enabling the development of liquid-crystal flat-panel displays (Volume 2, Chapter 3).

The MOSFET transistor, the basic component of microprocessors and memories, is based on the second type of field effect: the inverse field effect. Studies on this field effect began in 1960 (Chapter 6).

3.1.2. *Bell Labs studies*

In the summer of 1945, according to Ian Ross, President of Bell Labs from 1979 to 1997, Kelly (Director of Research, then President of Bell Labs from 1936 to 1959) set up a research group with the aims of "the fundamental study of semiconductors", concentrating on germanium and silicon, materials which were beginning to be well known; and in the long term "the creation of a solid state component" to replace the triodes (vacuum tubes) used as signal amplifiers for the transmission of information along telephone lines[1] and as switches in the electro-mechanical relays of ATT telephone exchanges (Ross 1997).

III-V compound semiconductor materials such as gallium arsenide (GaAs) were only discovered in 1951. They have not succeeded in dethroning silicon in its essential applications.

According to Shockley[2], even before the war, Kelly had in mind the idea of finding a solid component to replace triodes. Shockley, who had been involved in Brattain's studies of solid Cu/Cu_2O rectifiers, came up with the idea that it should be possible to obtain an amplifying effect similar to that obtained with a triode with a Cu/Cu_2O rectifier. This was an idea which he recorded in his laboratory notebook on December 29, 1939, in a note entitled "a semiconductor triode as amplifier" and renamed by Shockley "a Schottky-gate field-effect transistor" (Shockley 1976 p. 197).

The discovery of silicon N and P (described in Chapter 2) in 1940, and the control of silicon conductivity by doping, led Shockley to the idea of studying the effect of an electric field on silicon conductivity. So, from the end of the war, under

---

1 Triodes reached their apogee with the construction of the first major intercontinental submarine cable links, using submerged amplifiers. The first project was completed in 1956. But the future belonged to the transistorization and digitization of links.

2 "It occurred to me sometime around 1938 or 1939 that there should be a possibility of achieving the objective of electronic switching that Dr. Kelly had in mind using phenomena in solid-state physics rather than the vacuum-tube techniques that he had considered" (Shockley 1976, p. 597).

his impetus, studies at Bell Labs focused on the field effect[3], that is, the effect of an electric field on the electrical conductivity of an N or P semiconductor.

Numerous experiments on the direct field effect were carried out by Pearson as early as 1945 (reported in 1948). The experimental set-up consisted of a capacity made of a slice of fused quartz 75–100 μm (1 × 2 cm), covered on one side with a film of germanium N and a gold deposit (the gate) on the other (Figure 3.3). A high voltage applied to the gold deposit caused an increase in the current flowing through the germanium film (measured between two gold electrodes (source and drain) deposited on the semiconductor film). Measurements were made on film of germanium P, silicon N and copper oxide CuO_2[4]. These experiments showed only a very weak effect, contrary to theory (Shockley and Pearson 1948).

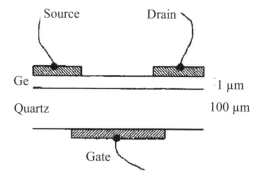

Figure 3.3. *Diagram of the device used by Pearson to demonstrate the field effect (Orton 2004, p. 75).*

3.1.2.1. *"Surface states": "surface traps"*

The relatively negative result of the field effect experiment was explained by Bardeen. It was assumed that the electrons accumulated under the surface of the N semiconductor by the electric field (Figure 3.2(b)) would be free to move like the electrons inside the crystal. In fact, the electrons accumulated on the surface were trapped (Figure 3.4). Surface-trapping states (or surface traps) (Figure 3.4(a)) were formed by the "dangling bonds" of the semiconductor surface (Figure 3.5(a)) and/or

3 Descriptions of the work carried out at Bell Labs by some of the major players who led to the invention of the transistor can be found in the articles by Pearson and Brattain (1955) and Brattain (1968).

4 With Cu_2O, an increase in relative conductivity of 0.11 at an electric field of 400,000 V/cm (V = 3,000 V) was achieved. "Although the modulation of 0.11 is not great, the useful output power is substantial. It is in principle operative as an amplifier. This is a moral victory", wrote Pearson in his laboratory work, as reported by Riordan and Hoddesson (1997, p. 143).

by impurities adsorbed on the surface. Electrons in the immediate vicinity of the surface lose part of their energy and "fall" into the energy band gap at an intermediate energy level called surface states (Figure 3.5(b)). They are temporarily trapped (Figure 3.4(b)) and cannot contribute to the electron flow to the surface and also to the expected amplifying effect (Bardeen 1950).

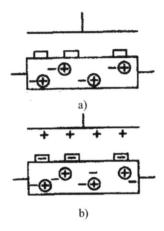

Figure 3.4. *Electron trapping: (a) surface traps; (b) trapped electrons in surface traps. Figure from Bardeen's laboratory notebook (Shockley 1976, p. 606)*

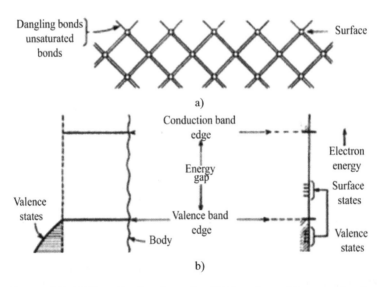

Figure 3.5. *(a) Dangling bonds on the {100} surface; (b) energy levels at the center of the band gap: surface states (Atalla 1959)*

3.1.2.2. *Demonstrating the direct field effect*

In order for the field effect to work, and for electrons to be free to move, these surface traps had to be blocked. Bardeen and Brattain proved the existence of these superficial traps in an experiment carried out on November 17, 1947, showing that the traps could be neutralized. At the suggestion of Robert Gibney, a physical chemist in the group, they showed that in the presence of a light source, by immersing the component (capacity), consisting of a metal electrode and a silicon P wafer in water (with a relative dielectric constant of 70–80), and applying a positive potential to the electrode, the ions in the solution attached themselves to the dangling bonds on the surface and neutralized the traps; the holes in the silicon P wafer were repelled from the surface. The electrons created by the photoelectric effect accumulated on the surface of the wafer (Figure 3.2(b)), their presence as free carriers being confirmed by measurement of the potential barrier created. This experiment was described by Shockley as follows: "this new finding was electrifying" (Shockley 1976 p. 606).

Based on Gibney's suggestion, Bardeen and Brattain (as early as November 20, 1947), tested a number of devices (Figures 3.6 and 3.7) for obtaining current amplification by direct field effect (majority carrier modulation).

The first device tested (Figure 3.6) consisted of a silicon P wafer, on whose surface a thin layer (one micron thick) of silicon N had been deposited by the CVD process and two circuits. The circuit (on the right) was made up of a "source" electrode (5) at positive potential, the thin N layer and the base (4). As the NP junction was under reverse polarity (see Chapter 4, Figure 4.3(a)), the current through the diode had to be very low. The circuit on the left was made up of a drop of electrolyte (12) (not reacting with the N layer) surrounding the insulated source electrode (5), and a ring electrode (13) placed in the drop of liquid. By varying the positive potential of the control electrode (13) via the source (10), and thus the electric field in the N-layer, and hence the conductivity of the layer, the current flowing in the right-hand circuit (source electrode (5) – base electrode (4)) spread out in the N-layer and could be modulated by the voltage applied to the electrode (13). With silicon, Bardeen and Brattain obtained significant power amplification, albeit at a low level (Bardeen 1947). Numerous traps in the polycrystalline N-silicon layer greatly reduced its conductivity.

On December 8, at Bardeen's suggestion, Brattain resumed tests on a device similar to the one shown in Figure 3.6, but with only one wafer of high-purity germanium N with large, high-purity crystals (Figure 3.7). By applying a negative potential to the source electrode (21) and the control electrode (25), placed in a drop of electrolyte (glycol borate), an unexpected and surprising amplifying effect was achieved. Voltage amplification by a factor of 2 and power amplification by a factor of 330 were observed (Riordan et al. 1999). However, the device did not amplify high frequencies. The frequency response was limited to 10 Hz due to the low mobility of the electrolyte ions.

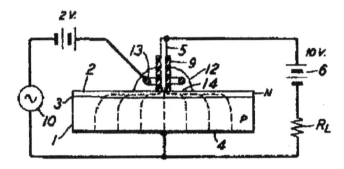

Figure 3.6. *Direct field effect: first experimental set-up used by Bardeen and Brattain (1950)*

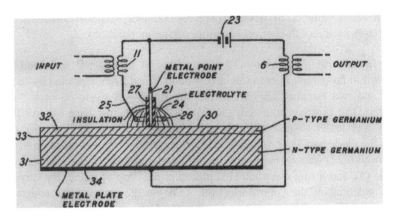

Figure 3.7. *Direct field effect: second experimental set-up used by Bardeen and Brattain (Brattain and Gibney 1950, p. 31)*

To explain this result, Bardeen[5] suggested that the very strong electric field created by the control electrode (25), raised to a negative potential, by repelling electrons from the surface layer of the N wafer, induced the formation of a P layer, an "inversion layer", on the surface of the N wafer (visible Figure 3.7) and thus a p-type surface conductivity. An inverse field effect would have been demonstrated.

---

5 "Bardeen suggests that the surface field is so strong that one is actually getting p-type conduction near the surface [...] and the negative potential on the grid is increasing the p-type or hole conduction", writes Brattain (1947).

By varying the negative potential of the control electrode, there was an increase in the conductivity of this surface layer; the current flowing in the right-hand circuit could be modulated, and thus a potential or power amplification was obtained (Brattain 1947).

In fact, the electrolyte (glycol borate), under negative potential, had produced, by electrolysis, an anodic oxide layer on the surface and an underlying P layer of very low thickness (1 μm) (Gibney 1951). The effect observed was therefore a direct field effect modulation of the conductivity of a thin P layer by a field of negative polarity (Figure 3.7).

By these experiments: "In other words, Bardeen and Brattain had demonstrated that the concept of the field-effect transistor was sound" (Seitz and Einspruch 1998, p. 167).

3.2. The germanium-based point-contact transistor

3.2.1. *The discovery of the germanium N point-contact transistor*

Based on these experiments, on a germanium N wafer covered with an electrolyte layer (glycol borate), which led to the formation of an anodic oxide layer, eliminating the electrolyte, Bardeen and Brattain deposited a gold film on this oxide layer to act as a control electrode (electrode 25 in Figure 3.7) and create a transverse electric field. A hole was drilled in the gold film to allow the "source" electrode to contact the wafer (electrode 21 in Figure 3.7). No effect was observed. The oxide layer had been removed during the block washing operation to eliminate the electrolyte, and the gold film, now in direct contact with the germanium plate, constituted a second electrode. The latter (the gold film) was then subjected to a positive potential and the tip electrode to a negative potential. Bardeen and Brattain, by moving the tip electrode closer to the edge of the hole in the gold film, found that a current flowed in the collector circuit, and that by varying the positive potential applied to the gold film, they varied the current flowing in the tip diode circuit, and thus the voltage across a resistor. The result was power amplification. Clearly, something new was happening.

Replacing the gold film with a second gold point electrode at a positive potential (Figure 3.8), Bardeen and Brattain observed a current circulation in the collector circuit, for a variation in voltage inducing a variation in current in the emitter circuit (by application of an AC signal), when the electrodes were sufficiently close (50 μm apart). A power gain of 4.5 for a current frequency of 1 kHz was obtained (Bardeen and Brattain 1948). "The transistor was discovered on December 16, 1947" (Seitz and Einspruch 1998, p. 168)[6].

Voltage and power amplifications of a factor of 100 were then obtained up to frequencies of 10 MHz with, depending on the configuration, current amplifications of a factor varying from 1 to 2; in later tests, this reached a factor of 3 (Bardeen and Brattain 1949).

These performances were achieved with germanium N with high purity and large crystals (absence of lattice defects traps), allowing a current to flow between the emitter and collector electrodes due to the absence of such traps (see Chapter 5, section 5.1.2).

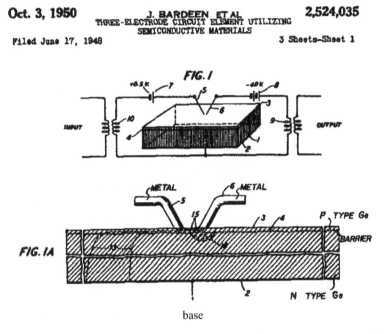

Figure 3.8. *The germanium N point-contact transistor. Figures show the two circuits and their power supplies (Bardeen and Brattain 1950)*

---

6 We quote a few sentences from Seitz and Einspruch's book: "they made a remarkable discovery"; "Bardeen and Brattain evidently had discovered far more than they had bargaining"; "the full importance of this partly chance discovery; a triode that employed an entirely new principle was born. It eventually came to be called the bipolar point-contact transistor" (Seitz and Einspruch 1998).

Bardeen and Brattain continued all their tests with high-purity germanium N, which explains the development of germanium transistors.

As soon as the discovery/invention by Bardeen and Brattain was released, in June 1948, a large number of companies began producing such components.

In July 1948, Jerome Kurshan of the RCA laboratories designed a transistor based on a germanium N point-contact diode (Burgess 2008).

IBM also embarked on similar research, starting with germanium point-contact diodes (Bashe et al. 1986). IBM succeeded in obtaining current-amplifying transistors with an electrode drawn from a copper alloy and soldered to the base by applying a high current. By the spring of 1950, IBM had made enough progress in the development of these transistors that production was launched, using point-contact diodes purchased from General Electric. IBM set about designing and building logic circuits. However, the lack of reliability and predictability meant that no IBM computer was ever built with these transistors.

3.2.2. Operation of the germanium-based N-tip transistor

The component, having demonstrated the amplifying effect, consists of two circuits of two point-contact diodes, one biased in the forward direction (the emitter circuit, on the left) and the other biased in the reverse direction (the collector circuit, on the right). The semiconductor is germanium N (Figure 3.9).

When the emitter and collector electrodes are separated from each other, currents flow through the two point-contact diodes independently.

In the emitter circuit, the current flows in the forward direction. In theory, for a metal-semiconductor N diode, the current flowing consists of a current of electrons from the base to the electrode, and a current of holes from the electrode to the base (Chapter 2, section 2.5, formulas [2.2] and [2.5]). To explain the amplifying effect in voltage and power, Bardeen and Brattain postulated that the current is composed mainly of holes "injected" into the base by the emitter electrode ("current from the emitter is composed in large part of holes"); the holes gradually undergo "electron-hole recombination" with the electrons in their path through the base to the negative electrode of the emitter circuit (see Chapter 1, section 1.2.5) (Bardeen and Brattain 1949).

A saturation current flows in the collector circuit, which is polarized in the opposite direction (blocked direction). This current is relatively high in the case of germanium, as shown in the Appendix in section 2.5.

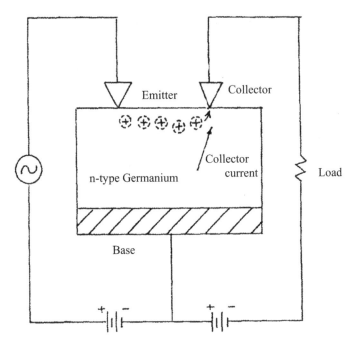

Figure 3.9. *Point-contact transistor: electrical circuit and current of holes "injected" by the emitter electrode and flowing to the collector electrode (Brattain 1968)*

When the two electrodes are close enough (Figure 3.9), the holes "injected" by the emitter into the base, attracted by the high negative potential of the collector electrode (30–40 V), can pass through the N-base between the two electrodes and reach the collector without recombining with electrons. The current of holes δI_C captured by the collector is only close to the hole current emitted by the δI_E emitter if the holes injected into the "base" reach the collector without falling into a trap; that is, without recombining with a conduction electron.

The voltage V or power amplification measured across the collector circuit resistor (load) R is due to the current of holes I "injected" by the emitter electrode and attracted by the collector electrode at negative potential (V = RI).

In addition, the holes arriving in the depletion zone of the collector electrode lower the potential barrier presented by the depletion zone, allowing an increase in the leakage current in the collector circuit. The amplifying effect of up to a factor of 3 is due to the increase in this reverse (leakage) electron current. The current of holes emitted by the emitter controls and amplifies the reverse current flowing in the collector circuit (Bardeen and Brattain 1949).

In a first interpretation, for Bardeen and Brattain, the anodic oxidation treatment of germanium N had created a P surface layer on the germanium N wafer (see section 3.1.2.2); the holes "injected" by the emitter electrode, attracted by the strong negative potential of the collector electrode (30–40 V), reached the collector electrode via this P layer.

However, the amplifying effect was also observed on an identical configuration in the absence of a surface layer P. The phenomenon was then interpreted as due to "the injection of holes by the electrode of the emitter circuit into the N-base and the circulation of holes in the N-base from the emitter electrode to the collector electrode".

Numerous experiments were then carried out to verify the proposed mechanism, namely the phenomenon of injection and circulation of holes in an N semiconductor (Haynes and Shockley 1949, 1951).

Shive's experiment, presented in Chapter 5, section 5.1.1.2 (Figure 5.2), demonstrated that there was indeed an injection of "holes" from the emitter and the circulation of holes (minority carriers) through the base of different polarity N to the collector.

The phenomenon of injection of holes (p minority carriers) by the emitter into the N base of a germanium-based diode N and the corresponding hole current are presented in Chapter 2, section 2.5.

The amplifying effect observed by Brattain and Bardeen on their "component" is the result of a combination of favorable factors: a semiconductor with a low forbidden bandwidth (0.66 eV), imposing a substantial, if not majority, current of minority carriers (holes) injected into the germanium N base; an N-doped low-purity germanium wafer with large crystals (practically mono-crystalline), between the two metal tips (50 μm apart); and with a "lifetime" of holes (minority carriers) injected by the emitter electrode long enough for them to reach the collector electrode (see Chapter 1, section 1.2.5 and Chapter 4, section 4.3.2).

The amplifying effect would not have been observed if the experiments had been carried out with silicon, with a bandgap of 1.12 V. The current of holes injected by the emitter electrode is very low (of the order of 10^{-6} times the electron current); it would have had no influence on the reverse electron current in the collector circuit.

Apart from the discovery of the transistor, that is, an amplifying component (for which Bardeen and Brattain were awarded the Nobel Prize), the second extremely important result was the identification of the mechanism of injection and circulation by diffusion of holes (minority carriers) in a semiconductor of the opposite type (n-type); this led to the invention and subsequent production of the bipolar transistor with NP junctions by Shockley.

The mechanism of injection and circulation of n or p "charge carriers" in a region of opposite type p or n is presented in Chapter 2, section 2.5 and Chapter 4, section 4.1.2.

3.2.3. *The point-contact transistor by Herbert Mataré and Heinrich Welker*

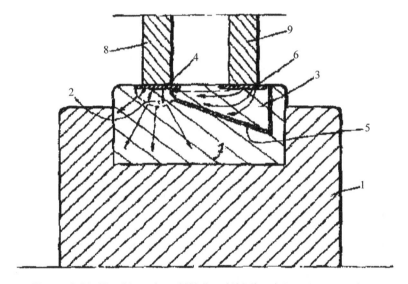

Figure 3.10. *The Mataré and Welker (1954) point-contact transistor*

In March 1946, Heinrich Welker[7] was invited to work with Herbert Mataré at a laboratory in France: *"compagnie des freins et signaux Westinghouse"* to develop and manufacture germanium diodes of the type produced by Sylvania Electric Products (IN34) in 1946 (Chapter 2, Figure 2.15). During the war, these two researchers had worked on the development of the point-contact diode as a radar component. In France, Mataré (1951) designed a rectifier (push-pull converter of the crystal type for ultra-short waves) as a radar component, consisting of a crystal and two closely spaced electrodes, a device for suppressing the oscillator noise in heterodyne receivers, which he patented in France in May 1947. Observation of the operation of this device led to the invention, announced in August 1948, of a germanium multi-electrode transistor called "Transistron", similar in design to the

7 Heinrich Welker: the father of III-V intermetallic semiconductors (Nobel Prize).

Bell Labs transistor. The components were made from high-purity germanium produced by Welker. The components consisted of two adjoining zones of N (1) and P (3) and two closely spaced electrodes (needles) in contact with the two zones, respectively (Figure 3.10). A French patent was filed on August 13, 1948, followed by a U.S. patent in 1949 (CHM 1948; Mataré and Welker 1954). Thousands of such components were manufactured in France for the telephone system.

3.3. The industrial development of the germanium N point-contact transistor

During the first months of 1948, Pfann developed a shrouded version of this transistor and manufactured over a hundred of them. "The production of the first tip transistors was a nightmare", according to Riordan and Hoddesson (1997, p. 157).

The availability of high-purity single crystals of germanium from 1950 onwards (see Chapter 4) enabled Bell Labs to solve the problems of uniformity and reliability of these transistors, albeit with a very high susceptibility to temperature variations (which would be a major cause of the development of silicon transistors imposed by the US army).

Point-contact transistors were developed between 1948 and 1951, and commercial production began in 1950 by ATT subsidiary Western Electric (Figure 3.11). Production lasted until the 1970s. They were used as switches in telephone exchanges.

Figure 3.11. *Point-contact transistor manufactured by Western Electric: base in germanium N emitter in 180 μm beryllium copper wire, collector in phosphor bronze wire (Ward 2015). For a color version of this figure, see www.iste.co.uk/vignes/silicon1.zip*

Numerous companies began manufacturing such transistors: Sylvania (GT-372), Raytheon (CK703) and General Electric (SX-4A and Z2). Texas Instruments manufactured such transistors (types 100, 101 102) in 1952, but soon stopped production.

As soon as Bardeen's article announcing the discovery of the point-contact transistor was published in 1948, General Electric began manufacturing such transistors, using the basic technologies (developed for the point-contact diode) (Chapter 2). Two types of transistors were proposed: an amplifier transistor (G11A/2N30) and a switching transistor (G11A/2N31) (Burgess 2008, 2011).

Figure 3.12. *Point-contact transistors manufactured by General Electric, G11-G11A (Burgess 2008, 2011). For a color version of this figure, see www.iste.co.uk/vignes/silicon1.zip*

RCA began building prototypes in November 1948 (TA150 batch of 200), presenting a cut-off frequency of 20 megacycles obtained by a distance between the tips of 1 mil (25 μm) for TV amplifiers. Commercial production began in May 1952 (Burgess 2008). Other companies began manufacturing such transistors (Westinghouse, CBS).

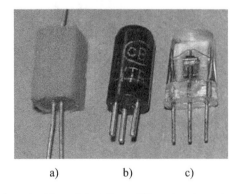

a)   b)   c)

Figure 3.13. *Point-contact transistors from different companies: (a) Westinghouse WX3347; (b) CBS PT-2A; (c) RCA TA-165 (Ward 2019). For a color version of this figure, see www.iste.co.uk/vignes/silicon1.zip*

By 1953, Telefunken, a Bell licensee, was producing and marketing point-contact transistors.

These transistors offered radio-frequency amplification performances up to 20 megacycles higher than those of the first bipolar transistors (Chapter 5), thanks to the possibility of realizing very small distances between electrodes.

In addition, as switching elements, they found applications in telephone exchange switches. Thousands of such components were manufactured in France for the telephone system.

In 1953, the first prototype computer using germanium point-contact transistors was built by Manchester University (92 transistors and 550 diodes).

The TRADIC phase one calculator, built by Bell Labs from 1951 to 1954, was the first transistorized digital calculator, executing additions and subtractions in 16 μs, and multiplications and divisions in less than 300 μs (Irvine 2001). It used 10,358 germanium N point-contact diodes, one-third of which were used for AND and OR logic circuits, and 684 point-contact transistors as amplifiers for the attenuated signals output by DRL (diode-resistor logic) circuits (Volume 2, Chapter 1). The logic circuits of the TRADIC 1, 2 and 3 computers were based on patents by John Felker of Bell Labs (Felker 1952).

The first prototype computer was manufactured in Japan in 1956, the ETL Mark 3, which featured 130 point-contact transistors and 1,800 germanium point-contact diodes.

3.4. References

Atalla, M.M. (1959). SF system: Power conversion. *Bell System Technical Journal*, 5, 749.

Bardeen, J. (1946). *Bell Labs Notebook.* AT&T Archives, Warren.

Bardeen, J. (1950). Three-electrode circuit element utilizing semiconductive materials. Patent, US2524033.

Bardeen, J. and Brattain, W.H. (1948). The transistor: A semiconductor triode. *Physical Review*, 74, 230–231.

Bardeen, J. and Brattain, W.H. (1949). Physical principles involved in transistor action. *Physical Review*, 75(8), 1208–1226.

Bardeen, J. and Brattain, W.H. (1950). Three-electrode circuit element utilizing semiconductive materials. Patent, US2524035.

Bashe, C.J., Johnson, L.R., Palmer, J.H., Pugh, E.W. (1986). *IBM's Early Computers*. The MIT Press, Cambridge [Online]. Available at: https://mitpress.mit.edu.

Brattain, W.H. (1947). *Bell Labs Notebook*. AT&T Archives, Warren.

Brattain, W.H. (1950). Three-electrode circuit element utilizing semiconductive materials. Patent, US2524034.

Brattain, W.H. (1968). Genesis of the transitor. *The Physics Teacher*, 3, 109–114.

Burgess, P.D. (2008). RCA transistor history. Transistor history [Online]. Available at: https://sites.google.com/site/transistorhistory/Home/us-semiconductor-manufacturers/rca-history.

Burgess, P.D. (2011). General Electric history. Transistor history [Online]. Available at: https://sites.google.com/site/transistorhistory/Home/us-semiconductor-manufacturers/general-electric-history.

CHM (1948). The European transistor invention. The silicon engine, timeline [Online]. Available at: www.computerhistory.org.

Felker, J.H. (1952). Regenerative transistor amplifier for digital computer applications. *Proceedings IRE*, 11, 1584–1591.

Gibney, R.B. (1951). Electrolytic surface treatment of germanium. Patent, US2560792.

Haynes, J.R. and Shockley, W. (1949). Investigation of hole injection in transistor action. *Physical Review*, 75(4), 691–844.

Haynes, J.R. and Shockley, W. (1951). The mobility and life of injected holes and electrons in germanium. *Physical Review*, 81(5), 835–844.

Irvine, M.M. (2001). Early digital computers at Bell Telephone laboratories. *IEEE Annals of the History of Computers*, 23(3), 22–42.

Mataré, H.F. (1951). Push-pull converter of the crystal type for ultra-short waves. Patent, US1252052.

Mataré, H.F. and Welker, H. (1954). Crystal device for controlling electric currents by means of a solid semiconductor. Patent, US2673948.

Orton, J.W. (2004). *The Story of Semiconductors*. Oxford University Press, Oxford.

Pearson, G.L. and Brattain, W.H. (1955). History of semiconductor research. *Proceedings of the IRE*, 43(12), 1794–1806.

Riordan, M. and Hoddeson, L. (1997). *Crystal Fire: The Invention of the Transistor and the Birth of the Information Age*. W.W. Norton & Company, New York.

Riordan, M., Hoddesson, L., Herring, C. (1999). The invention of the transistor. *Review of Modern Physics*, 71(2), 336–345.

Ross, I. (1997). The foundation of the silicon age. *Bell Labs Technical Journal Special Issue: The Transistor 50th Anniversary*, 2(4), 3–14.

Seitz, F. and Einspruch, N.G. (1998). *Electronic Genie: The Tangled History of Silicon.* University of Illinois Press, Illinois.

Shockley, W. (1976). The path to the conception of the junction transistor. *IEEE Transactions on Electron Devices*, 23(7), 597–620.

Shockley, W. and Pearson, G.L. (1948). Modulation of conductance of thin films of semi-conductor by surface charges. *Physical Review*, 74, 232.

Ward, J. (2015). History of transistors. Transistor Museum [Online]. Available at: http://semiconductormuseum.com/Museum_Index.htm.

Ward, J. (2019). History of transistors. Transistor Museum [Online]. Available at: http://semiconductormuseum.com/.

The PN Diode

The discovery of the rectifier effect presented by a "PN junction" by Russell Ohlin, 1940, was the founding act of William Shockley's 1949 invention of the bipolar transistor, a component consisting of two PN diodes placed side by side.

The PN diode is a current rectifier for low and high currents up to thousands of amperes. Due to its high switching speed, it is a component of logic circuits in computers and so-called electronic telephone exchanges. The PN diode is also one of the basic structures of microelectronics components: NPN bipolar transistors, memories and solar cells.

The current in a diode, made up of a current of electrons and a current of holes, is controlled by the generation/recombination of electron–hole pairs, themselves controlled by metallic impurities and lattice defects of the base material.

The development of purification and crystal growth processes for germanium and then silicon, thanks to the persistence of metallurgists, ensured the development of PN diodes and bipolar transistors. The purity (10N) and crystalline perfection achieved solved the problems of uniformity and reliability.

The development of the zone fusion purification process ensured germanium's dominance in the 1950s as the base material for NPN bipolar transistors.

Germanium-based PN diodes are "leaky switches"; the higher the temperature, the more leaky they become. It was this property which, under pressure from the military, led to the development of processes for the purification and crystal growth of silicon. The invention of the floating-zone process made it possible to achieve the purification and crystal growth that would eventually, albeit slowly, establish silicon as the basic material for transistors.

This chapter presents:

– the discovery of the PN diode;

– the PN diode operation;

– the PN diode functions;

– the physical basis of PN diode operation (see the Appendix);

– the effect of impurities on PN diode characteristics;

– the development of PN diodes;

– the purification, doping and manufacturing processes for "electronic" silicon and germanium single crystals.

4.1. PN diode operation and functions

The PN diode is a component consisting of two semiconductor regions, N and P, in abrupt planar contact, one p-doped and the other n-doped, *of the same semiconductor single crystal.*

4.1.1. *The discovery of the rectifier effect of the silicon-based PN diode*

The first PN diode was extracted from a silicon ingot slowly solidified in helium, produced at Bell Labs in 1940 by Jack Scaff's team at the request of Russell Ohl (Figure 4.1). On a longitudinal slice of the ingot, Russell Ohl[1] discovered that this wafer exhibited the rectifying effect of point-contact diodes. He also discovered that this wafer exhibited the photovoltaic effect (see Volume 2, Chapter 4). Ohl and Scaff showed that the formation of the PN junction must be the result of impurity segregation during solidification (Chapter 2, section 2.3.1) (Scaff 1946, 1970).

When the remarkable rectifying and photovoltaic properties of this ingot were brought to the attention of Bell Labs Director Mervin Kelly decided that this discovery was of great value to the electronics industry, and that absolute secrecy should be maintained until further study revealed its full power: "it was too important a breakthrough to bruit about" (Seitz and Einspruch 1998, p. 157). The studies were resumed in 1945 (Chapter 3, section 3.1.2). The embargo concerned the PN junction and its rectifying and photoelectric properties, not the discovery of P and N.

---

1 "An unsung hero of semiconductor history: Russell Ohl the inventor of the pn junction. Yet the name of its inventor remains largely unknown to the many who benefit from his crucial break-through" (Riordan and Hoddesson 1997, p. 208).

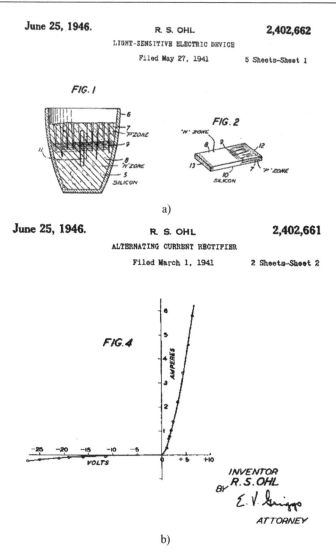

Figure 4.1. *(a) The silicon-based PN junction (Figure 2) extracted from the unidirectionally solidified silicon ingot (Figure 1) (Ohl 1946a); (b) the characteristic curve of the diode extracted from the silicon ingot (Figure 4) (Ohl 1946b)*

4.1.2. *PN diode operation*

PN diode operation is controlled by the generation/recombination of electron–hole pairs (Chapter 1, Figure 1.8), themselves controlled by metallic impurities and lattice defects of the base material.

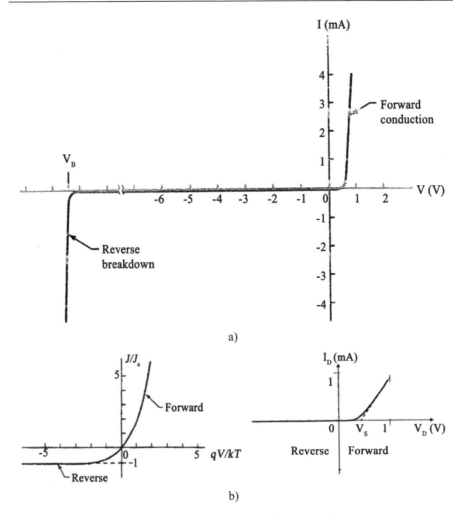

Figure 4.2. *(a) Characteristic curve of a silicon PN diode; (b) ideal characteristic curve. Above a threshold voltage Vs, the current varies linearly (Sze 2002, p. 88)*

The DC current–voltage characteristic curve (Figures 4.1 and 4.2(a)) has the same appearance as that of a point-contact diode (Chapter 2, Figure 2.1). As with a point-contact diode (or a metal-semiconductor (N or P) diode), the rectifying effect is due to the formation of depletion layers, on either side of the junction (Figures 4.3 and 4.4), depopulated of free charge carriers (electrons or holes) forming a potential barrier V_{bi}, whose overall thickness decreases with forward polarization and increases with reverse polarization (Figures 4.4 and 4.21).

The current per unit area I (or J) varies exponentially with the applied voltage V (Figure 4.2(b)):

$$I = I_S \cdot \{\exp(eV/kT) - 1\}$$
(section 4.3, equation [4.13])

In reverse polarity, the current is the saturation current I_S, also known as leakage current.

In direct polarity, for a voltage $V > V_s$ ("threshold voltage") $\sim V_{bi}$ (section 4.3, formula [4.1a and b]), of the order of 0.7 V for silicon and 0.3 V for germanium, the potential barrier becomes very low and the current is limited only by the resistances of the N and P regions. The regime becomes ohmic, and current increases linearly with voltage (Figure 4.2(b)).

In reverse polarity, above a certain voltage, known as the breakdown potential (or peak reverse voltage), the electric field generated by the high voltage produces an avalanche of electrons that destroys the diode.

When an N-region and a P-region are joined, electrons from the N-region move into the P-region, and holes move from the P-region into the N-region. Two layers are formed, joined together and depopulated of free charge carriers – "depletion layers" – carriers of fixed charges: the ions. The potential barrier established, V_{bi}, prevents further transfer of majority carriers from one region to the other (Figures 4.20 and 4.21).

Under direct polarity (forward) (Figure 4.3(a)), when a DC voltage is applied across a PN diode, with the positive pole connected to the P region and the negative pole to the N region (hence the names Si-P and Si-N), the potential barrier is lowered and becomes $(V_{bi} - V_F)$ (Figure 4.21(b)); a current I_F flows from region P to region N: sum of a hole current I_P from region P to region N and an electron current I_n from region N to region P.

The steps involved in moving holes and electrons under direct polarity are shown in Figure 4.3(a):

$$J_F = J_N \text{ (steps } 3\to 1, 1\to 1', 1'\to 1'', 1''\to 4) + J_p \text{ (steps } 4\to 2, 2\to 2', 2'\to 2'', 2''\to 3)$$

Electrons are injected at the negative electrode (in 3). Under the action of the electric field, located in the depletion layer, electrons from the N region move into the P region (step $1\to 1'$). They become minority carriers. They "diffuse" into the P region (see section 4.3.2, formula [4.6]), gradually recombining with holes in the P region (in $1''$). This is the recombination of electrons with holes in the P region, a function of the "lifetime" (Chapter 1, section 1.2.5.2) of the minority carriers that controls the electron current J_n leaving the N region and entering the P region.

At the positive electrode, holes are injected into the P region (in fact, an electron–hole pair is generated; at 4, an electron breaks free from a covalent bond to leave the diode, creating a new hole). The injected holes (at 4) cross the P region, then the depletion layer ($2 \rightarrow 2'$) and "diffuse" into the N region, gradually recombining with electrons from the N region (in $2''$). It is the recombination of holes with electrons in the N-region that controls the hole current J_p leaving the P region and passing into the N region.

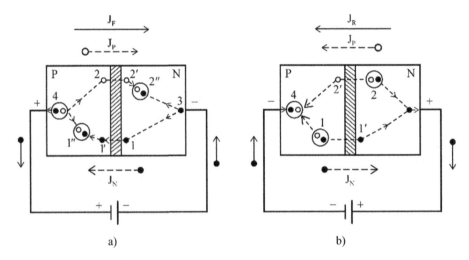

Figure 4.3. *Electron and hole currents in the PN diode:*
(a) with direct polarity; (b) with reverse polarity

Electron–hole recombinations (in $1''$ and $2''$), in the N and P regions, occur on the "traps" (consisting of impurities and lattice defects present in each region) (Chapter 1, section 1.2.5, Figure 1.8). Depending on the density of these traps, these recombinations will occur progressively over a greater or lesser distance from the boundaries of the depletion layer until the minority carriers disappear. This distance is called the "diffusion length" ($L_{p(N)}$ or $L_{n(P)}$). The greater the diffusion length, the longer the minority carrier can travel without being absorbed, and vice versa for a short diffusion length (Figure 4.4). The hatched areas represent the "charges" (quantities) of minority carriers in each region.

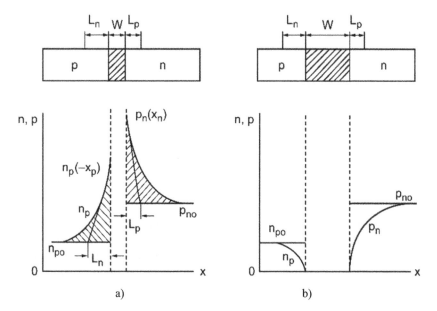

Figure 4.4. *Concentration profiles of minority carriers in each region: (a) under direct polarization; (b) under reverse polarity (Sze 2002, p. 107)*

The electron and hole currents are independent. The expression for the electron current in the P region is (see section 4.3.2):

$$I_{n(P)} = en_i^2 \, (D_{n(P)}/N_A \, L_{n(P)}). \, \{exp(eV/kT) - 1\}$$

(section 4.3, formula [4.9a]), where the exponential term is the electron concentration at the edge of the region n_p (-x_p), a function of the applied potential, and where $L_{(n)P}$, the "diffusion length" shown in Figure 4.4(a), is a function of the "lifetime" $\tau_{n(P)}$ of the minority carriers:

$$L_{n(P)} = (D_{n(P)} \cdot \tau_{n(P)})^{1/2}$$

(section 4.3, formula [4.10a])

The shorter the "lifetime" of the electrons in the p-region, the shorter the diffusion length; and the stronger the concentration gradient, the higher the current $I_{n(P)}$.

Similarly, the expression for the hole current in the N region is:

$$I_{p(N)} = en_i^2 \, (D_{p(N)}/N_D \, L_{p(N)}) \cdot \{exp(eV/kT) - 1\}$$

(section 4.3, [4.9b]), where:

$$L_{p(N)} = (D_{p(N)} \cdot \tau_{p(N)})^{1/2}$$

(section 4.2, formula [4.10b]), and $D_{p(N)}$ is the diffusion coefficient of holes in the N region.

The sum of currents $I_{n(P)} + I_{p(N)} = I_{Forward}$ is given by formula [4.13].

Lifetimes are equal: $\tau_{n(P)} = \tau_{p(N)} = \tau_m$, the concentration of recombination centers (traps) being the same throughout the single crystal making up the diode, and the capture and emission coefficients of electrons and holes being equal.

The "diffusion length" is the quantity we aim to maximize (in the case of solar cells and bipolar transistors acting as amplifiers) or to minimize (in the case of diodes and bipolar transistors acting as switches). It is controlled by the phenomenon of electron–hole recombination, itself controlled by "impurities" (section 4.1.4).

In reverse polarization, "blocked direction", with the negative pole of the battery connected to the P region (Figure 4.3(b)), under an applied potential V_R, the potential barrier increases: $V_{bi} + V_R$ (Figure 4.21(c)). The width of the depletion layer increases. The negative electrode attracts holes from the P region and the positive electrode attracts electrons from the N region. No current is allowed to flow. In fact, a "saturation current" J_R flows from the N region to the P region. A hole current generated in the N region (at 2) flows through the depletion layer $2 \rightarrow 2'$, then through the P region up to the negative electrode. The holes are absorbed by the electrons injected by the negative electrode through the electron–hole recombination process (in 3). Conversely, a current of electrons generated in the P region (at 1) crosses the depletion layer $(1 \rightarrow 1')$, then the N region. Electrons are extracted by the positive electrode.

A second current source is due to the generation of electron–hole pairs in the depletion layer. It is presented in the Appendix, section 4.3 (formulas [4.15] to [4.17]).

For a silicon diode, at ordinary temperature, the saturation current I_S is very low, of the order of 10^{-11} A/cm^2, due to the low intrinsic electron density n_i (Chapter 1, Table 1.1 and formula [1.1]).

On the other hand, the current due to the generation of electron–hole pairs in the depletion layer is of the order of $J_{gen} = 10^{-8}$ A/cm^2 for an applied reverse potential of 4 V.

For a germanium diode, in the blocked direction, the current is essentially the saturation current I_S. It is proportional to the square of the intrinsic carrier concentration n_i and therefore to the exponential of the bandgap $E_G = 0.67$ eV. The concentration of intrinsic carriers n_i is high, 2.8×10^{13}/cm^3 (Chapter 1, Table 1.1), so the current in the blocked direction is relatively high.

The variation of this saturation current with temperature is much more significant for germanium than for silicon, as n_i increases sharply with temperature, due to a much lower energy bandgap E_g (Chapter 1, Figure 1.5). This produces "leaky switches".

Saturation current (reverse current) is 1,000 times lower for silicon than for germanium. Temperatures of up to 150°C are tolerable for silicon diodes, whereas germanium diodes cannot withstand temperatures above 75°C.

4.1.3. *PN diode functions*

In logic circuits, the diode is a "switching" component; in analog circuits, the diode allows weak currents to pass up to a certain cut-off frequency.

The elementary circuit for both operating modes consists of a diode and a load resistor R_D (Figure 4.5(a)), and the operating point Q of the diode is determined by the intersection of the diode characteristic curve and the load line of the external circuit:

$$V_D = V_{DD} - R\ I_{extD}$$

For simplicity of presentation, we consider a diode of the type P$^+$ N, where the current is essentially a minority carrier current p (holes) in the N region, $I_{p(N)}$ (formula [4.9b] and Figures 4.3(a) and 4.4).

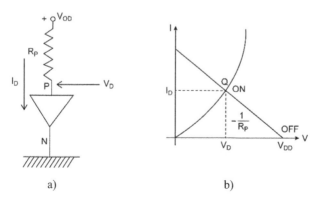

a)   b)

Figure 4.5. *(a) Elementary diode-resistor circuit with load R_D; (b) operating point Q (I_D/V_D) of the diode*

The diode can be modeled by an "RC" circuit[2] , like the point-contact diode (Figure 2.5) where the capacitance C is the sum of the depletion layer capacitance and the diffusion capacitance in the N region:

$$C = (Cj + C_D)$$

The capacity of the depletion layer is lower than the diffusion capacity. The diffusion capacity C_D is defined as the ratio between the "charge" Q_p (minority carrier storage) – the quantity (number) of holes (minority carriers) present in the N region, represented by the hatched area in the N region (Figure 4.4(a)) and the operating point potential V_D:

$$C_D = Q_P/V_D$$

The charge "storage time" is equal to the ratio of the charge Q_P to the current in the diode $I_{Pn} = I_D$. It is shown to be equal to the "lifetime$_P$" $\tau_{p(N)}$ of the p minority carriers in the N region:

$$Q_P = I_D \cdot \tau_{p(N)}$$

The shorter the "lifetime" $\tau_{p(N)}$ of the minority carriers (formulas [4.7] and [4.10]), the smaller the quantity of Q_P minority carriers present in the N region at a given time, the smaller the storage time of the minority carriers.

The resistance R is the slope of the characteristic curve around the operating point I_D, V_D (Figure 4.5(b)):

$$R = dV_D/dI_D$$

We show that the time constant of the RC is equal to the storage time and therefore to the lifetime of the minority carriers in the N region:

$$\tau_{RC} = \tau_{p(N)}$$

4.1.3.1. *The PN diode switching component*

Switching times are the times required to switch from ON to OFF (turn-off time) and vice versa (turn-on time).

The time it takes to go from OFF to ON (turn-on time) is the charge storage time τ_{RC} of the charge Q, equal to three times the RC circuit time constant (see Figure 5.8):

---

2 The formulas quoted can be found in the following monographs: (Sze 2002, p. 114; Hu 2009, p. 89).

$$\tau_{ON/OFF} = 3\ \tau_{RC} = 3\tau_{p(N)}$$

The turn-off time is the time required to eliminate the minority carriers (holes) present in the N region in the ON state by recombination with the majority carriers. It is of the order of magnitude of the "lifetime" of the minority carriers:

$$\tau(turn\text{-}off) = \alpha \cdot \tau_{p(N)} \text{ where } \alpha < 1$$

The expression de α can be found in the monographs.

4.1.3.2. *The PN diode as a component of analog circuits*

In weak-signal processing circuits featuring diodes, the diode operates around a "bias point" I_D, V_D, defined as the intersection of the diode characteristic curve and the external circuit load line $V_D = V_{DD} - R_{ext}\ I_D$. A weak (sinusoidal) signal current flows through the diode, superimposed on the bias point current $I_D - V_D$ (Figure 4.6).

Variations in diode voltage due to the signal to be processed generate current variations around the mean value.

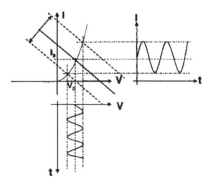

Figure 4.6. *PN diode. Conversion of weak signals (Koeniguer n.d., p. 7)*

The diode cut-off frequency of the diode, the maximum permissible weak-signal current frequency, is:

$$f_T = 1/2\pi\ \tau_{p(N)}$$

where $\tau_{p(N)}$ is the lifetime of minority carriers in the N region for a P+N diode.

4.1.4. *Action of dopants and impurities on the "lifetime" of minority carriers*

The action of defects, dopants and impurities is discussed in detail in two publications (Newman 1982; Sopori 1999).

4.1.4.1. *Minority carrier lifetime as a function of doping*

The electron–hole pair generation/recombination mechanism was presented in Chapter 1, section 1.2.5.

Figure 4.7(a) shows the lifetimes and diffusion lengths of a hole injected into an N-region, and Figure 4.7(b) an electron injected into a P-region as a function of the concentration of the respective dopants in these zones, for silicon purified by the floating zone (FZ) process (Figure 4.17), giving the highest achievable purity. For this silicon, this lifetime varies from 10^{-5} to 10^{-3} s for donor or acceptor concentrations ranging from $10^{18}/cm^3$ to $10^{13}/cm^3$. The "long" lifetime of electrons in a P region at low donor density is a feature that explains the (almost exclusive) use of silicon P as the basis for solar cell diodes (see Volume 2, Chapter 4). Conversely, for switching diodes, we look for the shortest lifetime.

Electron–hole generation/recombination takes place in two stages, each of which requires only small amounts of energy (according to the Shockley–Read–Hall mechanism) via a center (trap) that occupies an intermediate energy level in the energy band gap (localized levels).

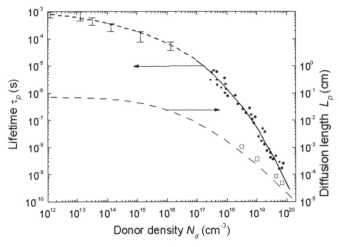

a)

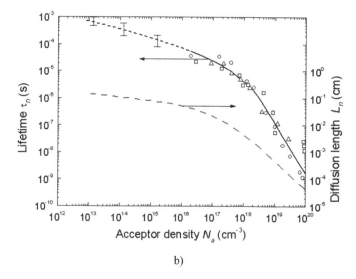

b)

Figure 4.7. *Lifetimes and diffusion lengths (a) for "holes" in an N region; (b) for electrons in a P region, in silicon. Variations with respective N_D and N_A dopant concentration (Meroli 2012). For a color version of this figure, see www.iste.co.uk/vignes/silicon1.zip*

Impurity atoms (strictly speaking, excluding dopants) and lattice defects (grain boundaries and dislocations) constitute such centers (traps) (see Chapter 1, Figure 1.8).

The lifetime (Chapter 1, section 1.2.4, formula [1.7] and section 1.2.5) of minority carriers in each region is inversely proportional to the concentration of electron–hole pair generation-recombination centers $N_{T(P)}$:

$$\tau_m \approx 1/N_{T(P)}$$

By minimizing the recombination-generation centers of electron–hole pairs and therefore impurities and lattice defects, we increase the lifetime of minority carriers.

4.1.4.2. Action of metallic impurities

Transition metals (Fe, Ni, Cr, Co) and noble metals (Cu, Ag, Au, Pt) have very low solubility in silicon. They dissolve in the crystal in either interstitial or substitutional positions. Gold is present in both forms. Iron exists in two forms: either interstitially or forming complexes with boron (Fe-B) or oxygen.

Iron, after being brought into solution at high temperature during heat treatment in manufacturing operations, on cooling, diffuses to interstitial sites; here it forms intermetallic compounds with silicon $FeSi_2$. The presence of $FeSi_2$ precipitate particles is totally undesirable, as they can form short circuits across a PN junction, and for MOSFET transistors (Chapter 6) cause gate oxide deterioration.

The outer electrons of transition metals occupy low energy levels in the energy band gap (Chapter 1, Figure 1.6). These require high energies to become free charge carriers.

Silicon conductivity is therefore only affected by the presence of these elements for concentrations above $10^{15}/cm^3$ (i.e. 20 ppba or 40 ppbwt).

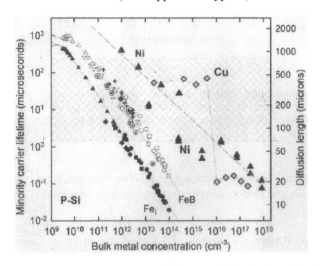

Figure 4.8. *Effect of Fe, Ni and Cu on lifetime and diffusion length of electrons (minority carriers) in boron-doped P single-crystal silicon for dopant concentrations between 10^{13} and $10^{16}/cm^3$ (Istratov et al. 2006)*

On the other hand, these transition metals, their complexes and precipitates either constitute electron–hole recombination centers (traps) that reduce the lifetime of minority carriers (Figure 4.8) (iron concentrations must be $<10^{11}/cm^3$ for the lifetime of N- or P-doped silicon not to be reduced), or electron–hole pair generation centers generating high leakage currents in reverse polarity, thus having a detrimental effect (see section 4.3, formulas [4.15] to [4.17]) (Weber 1983).

Gold is a more efficient center of electron–hole pair recombination than the transition metals (Figure 4.9). Its free electrons occupy an energy level close to the middle of the energy band gap (Chapter 1, Figure 1.6). For fast-switching devices, gold is added to minimize minority carrier lifetimes (Collins 1957).

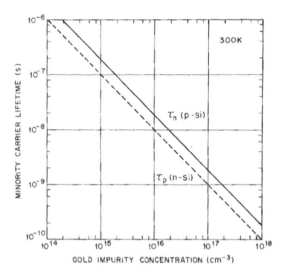

Figure 4.9. *Effect of gold on the lifetime of n and p minority carriers in the P and N regions (Collins 1957)*

If the concentration of metallic impurities in electronic silicon can be reduced to very low levels due to their low solubility and very low liquid/solid partition coefficient (segregation), such metallic impurities, iron and copper (elements called deathnium until they were identified), are introduced during manufacturing operations by contact with equipment; these penetrate (diffuse) into the material during heat treatment in certain manufacturing stages. The result is performance degradation due to contamination.

NOTE. – The problem of impurities in solar cells is dealt with in Volume 2, Chapter 4.

4.1.4.3. *Action of oxygen and carbon*

Oxygen is incorporated into the silicon in the Czochralski (CZ) single crystal pulling operation (presented in section 4.2.3.1) as a result of the dissolution of the silica in the crucible. Oxygen is present at concentrations of the order of 20–40 ppma (1–2×10^{18} atoms/cm^3). For concentrations close to the solubility limit, it occupies interstitial positions in the silicon crystal, forming two covalent bonds with two Si-O-Si silicon atoms, and is neutral in this configuration with regards to conductivity.

The oxygen forms SiO$_4$ complexes (clusters) which, after heat treatment at 430–450°C, precipitate out on cooling and act as electron donors (but by what mechanism is not known). These complexes are unstable at temperatures >500°C but can reform during manufacturing operations (diffusion treatments).

Oxygen can also form SiO_2 precipitates after high-temperature heat treatment (900–1,000°C). These precipitates act as "traps", electron–hole recombination centers contributing to the leakage current PN diodes in reverse polarity. Oxygen also acts as a gettering agent for metallic impurities. In addition, it pinches dislocations, enhancing the mechanical properties of the crystal. Oxygen precipitates on stacking faults, making silicon more resistant to the thermal stresses that occur during the many thermal cycles of manufacturing.

Carbon is present in concentrations from 10^{16} up to $3 \times 10^{17}/cm^3$ (i.e. 6 ppma, 6,000 ppba or 2,500 ppbwt), the solubility limit. It comes from the graphite components of the crystal pulling machine. It occupies lattice positions and does not act as a trap, neither as a donor nor as an acceptor.

4.1.5. *The development of PN diodes*

The PN diode is a current rectifier and, thanks to its high switching speed (Ciccolalla 1962), a component of logic circuits and an element of bipolar transistors (Chapter 5). It is also a component of solar cells (Volume 2, Chapter 4).

The operation of a PN diode, and therefore of all PN-junction transistors, is based on the following three characteristics:

– the hyper-purity of the base material to eliminate "traps";

– crystalline perfection of the basic structure (wafers) to eliminate lattice defects;

– doping control.

4.1.5.1. *The germanium PN diode*

The first germanium PN diode was manufactured in early 1950 using the CZ single-crystal pulling process by Bell Labs researchers (Leamy and Wernick 1997). Gordon Teal modified the single-crystal pulling system (section 4.2.4) to allow the controlled introduction of a dopant into the growing crystal (pill dropping). In the early months of 1950, Sparks began working closely with Teal and Buehler to produce PN junctions from a single-crystal germanium N in a Ga-doped P-type liquid bath (Teal 1955).

The first bipolar transistor with NPN junctions based on a germanium single crystal was produced by Teal in 1950 (April 10), directly using the CZ double doping grown-junction transistor process (presented in Chapter 5, section 5.2.1.1) (Shockley et al. 1951; Little and Teal 1954).

Single-crystal germanium was adopted as the material for diodes and transistors in the early 1950s.

In addition to the purification of germanium and production of germanium single crystals, a crucial development was the alloy junction process used by General Electric to manufacture diodes and transistors (Saby 1952) (described in Chapter 5, section 5.2.1). In 1951, this process was adopted by Radio Corporation of America (RCA) for the development of germanium PN diodes and PNP transistors, and subsequently by many other companies.

The relative ease of purification, single-crystal growth, controlled doping and the reliability of germanium PN diodes and PNP bipolar transistors were the key factors behind germanium's dominance from 1950 to 1965; this was mainly in the form of point-contact diodes (Chapter 2), PN junction diodes and PNP bipolar transistors.

Furthermore, we must add the following factors: the first is the relative ease of manufacturing a PN diode compared with a PNP transistor. For a long time, the transistor suffered from low yields and unreliable performance, which explained the longevity of diodes. Second, the diode was a much faster electronic switch than the NPN transistor.

Germanium PN diodes were used in the 1950s in some calculators (IBM 608-1957).

Germanium PN diodes conduct electricity at low voltages of 0.3 V. This makes them much more sensitive to weak signals than silicon diodes. For this reason, germanium diodes in weak signal processing circuits dominate in radios.

4.1.5.2. The silicon PN diode

During his last two years at Bell Labs (1951–1952), Gordon Teal with the help of Ernie Buehler (Teal 1952), sensing the future of silicon, succeeded in manufacturing silicon single crystals using the CZ process (Figure 4.15). Gordon Teal was familiar with this material, having manufactured silicon diodes during the war and observed its high-temperature stability characteristics (Teal 1976).

In 1952, although the characteristics of the first single crystals were mediocre due to the presence of relatively high levels of oxygen, the first silicon PN diode was manufactured by Pearson, using the CZ process used to produce the first germanium diodes by Pearson and Sawyer (1952) of Bell Labs. This diode proved superior to germanium diodes, with a leakage current 1,000 times lower, and could operate at temperatures up to 150°C. "Immediately silicon became a rival to germanium" (Leamy and Wernick 1997).

Given the difficulties of obtaining "electronic" silicon, silicon very gradually regained its pre-eminent position as a transistor material, under pressure from the military, who were practically the only customers at the time – particularly for the temperature resistance of silicon diodes and transistors up to around 150°C.

The ultimate supremacy of silicon over germanium, for PN diodes and NPN bipolar transistors, is due to two characteristics that derive directly from the respective values of the energy bandgap of these elements. The reverse current is 1,000 times lower for silicon than for germanium. Temperatures of up to 150°C are tolerable for silicon diodes, whereas germanium diodes cannot withstand temperatures above 75°C.

4.2. Electronic germanium and silicon production

It is to Gordon Teal of Bell Labs that we owe, from the discovery of the point contact transistor in December 1947, the recommendation to purify the material and to use single crystal as the base material of transistors. We also owe the development of the CZ pulling method to obtain single crystals of germanium initially, then silicon, the development of processes to obtain the greatest purity of the base material and the development of the doping process, for germanium and then silicon, which led to the production of the first bipolar transistor in germanium. This work was carried out covertly, without the authorization or approval of those in charge, particularly William Shockley[3].

Teal was the right man at the right place.

4.2.1. Electronic germanium production

In 1950, germanium was adopted as the material for PN diodes and NPN transistors.

4.2.1.1. Physical purification of germanium

Chemical purification processes for germanium were presented in Chapter 2, section 2.3.2.1. However, purification by chemical methods could not achieve sufficiently low purities <10^{14} a/cm^3.

---

3 "It can certainly be said, however, that the availability of such pure and perfect single crystals as we have in present-day silicon and germanium amounts to a major revolution in the physics of solids" (Pearson and Brattain 1955).

In 1948, General Electric developed the Bridgman process for purification by slow, progressive solidification of a germanium bath (Figure 4.10); as impurities accumulated in the last fraction of the ingot solidified, 80% of the ingot was substantially purified.

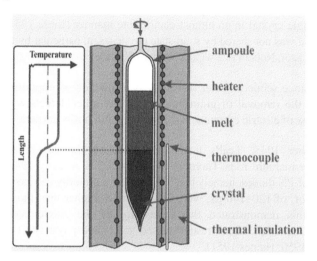

Figure 4.10. *Bridgman process. For a color version of this figure, see www.iste.co.uk/vignes/silicon1.zip*

4.2.1.2. *Production and doping of single-crystal germanium*

Gordon Teal had a special interest in germanium since completing his thesis on the properties of germanium at Brown University in the early 1930s, under the supervision of Professor Charles A. Kraus – a germanium specialist. In the early 1930s, recruited by the chemistry department of Bell Labs, he was involved in the preparation, testing and examination of electronic materials, including germanium, and developed the chemical vapor deposition (CVD) process for depositing germanium films by thermal decomposition of germanium hydride GeH_4 for the manufacture of telecommunication tube components.

From February 1942 onward, Gordon Teal initiated research into germanium and produced the first germanium rectifiers using the thermal decomposition of germanium chloride (gas digermane) (from Brown University) on tantalum filaments at high temperature. It was this work, presented to Professor Lark-Horovitz of Purdue University, which prompted him to conduct work during the Second World War on germanium as a material for germanium rectifiers, the industrial development of which was taken over by Bell Labs towards the end of the war. During the war, Bell Labs concentrated its efforts on silicon for radar receiver diodes, and it was Gordon

Teal who carried out the silicon deposition tests on a tantalum filament by thermal decomposition of SiCl$_4$ with hydrogen, from which silicon rectifiers were made (Storks and Teal 1948; Teal 1976).

Gordon Teal, in early 1948, with Little's support (Teal 1976), produced his first germanium single crystal in an almost clandestine manner (Little 1954)[4]. Interest in the single crystal was not shared by solid-state physicists, in particular by the inventor of the bipolar transistor, Nobel Prize winner William Shockley[5].

For Teal, since semiconductor resistivity is a function of impurities and crystal imperfections, the removal of grain boundaries and other defects was intended to improve the flow of electric current and the lifetime of minority carriers.

In September 1948, Teal's team[6] produced, by the CZ process, the first large-crystal germanium rods. Haynes' results at 300 K in early 1949 of the basic characteristic of PN diodes, namely the "lifetime" of n minority carriers injected into a P semiconductor, of 120–140 μs, were 20–100 times higher than those obtained on polycrystals; this demonstrated the superiority of the single crystal over the polycrystalline material. These results forced solid-state physicists to change their position (Teal 1950; Haines 1951).

In December 1948, Jack Morton, who headed the component development group, took Gordon Teal seriously and allowed him to install his single-crystal pulling equipment in his laboratory and further develop the technique. The year 1949 was devoted to purifying germanium by the fractioned crystallization method and producing single crystals of purified germanium. The resistivity of single-crystal germanium reached 45 Ω·cm (Chapter 1, Figure 1.13).

---

4 "On October 1, 1948, we had completed our crude machine in John's (Little) lab in New York city and pulled our first single crystals of germanium. We did this without getting anyone's permis-sion or approval and acted only on our personal ideas" (Teal 1976). "Teal focused on crystal-growth techniques because of his belief that single crystals would make better transistors, a view that was not prevalent at that time" (Leamy and Wernick 1997).

5 "Utilization of single crystal of germanium for the transistor was a very controversial matter at that time" (Huff 2002). "Teal tried to convince Shockley of this critical advantage, but Shockley ignored his suggestion. [...] Shockley and most of his colleagues initially argued vociferously for a polycrystalline technology: single crystals would be too expensive to be accepted by the produc-tion-line colleagues" (Riordan et al. 1999).

6 The CZ process, developed by Czochralski in 1917, was used to produce single crystals of alkaline chlorides (Kyropoulos 1926) and metals (tin and zinc) in 1929 by Hoyem and Tyndall (1929).

While the electron mobility in polycrystalline germanium in 1945 was 1,100 cm^2/V·s, in 1952, with germanium purified (by successive recrystallizations) and with a single-crystal structure, electron mobility reached 3,800 cm^2/V·s and hole mobility 1,700 cm^2/V·s at 300 K (see Chapter 1, Table 1.1).

Nevertheless, as reported by Jack Scaff then Director of the Materials Research Laboratory at Bell Labs, the variability of results was very great. In 1951, an interdisciplinary group of chemists, metallurgists, physicists and engineers was formed to tackle this problem. Methods for purifying germanium and then silicon were developed (Scaff 1970).

4.2.1.3. Purification of germanium by the zone melting process

Pfann had developed the "zone melting" (ZM) process to produce single crystals of lead-antimony alloys. In 1951, the germanium produced by this process reached a purity >99.99999999% (10N) (Pfann 1952, 1956, 1966). The purity achieved by this process was unprecedented in the history of materials processing. The availability of germanium single crystals of very high purity enabled the group led by Morton to solve the difficult problems of diode uniformity and fiability.

This process also made it possible to obtain doped germanium.

On the other hand, using the process for silicon purification proved problematic, as the silicon reacted with the quartz pods.

All the companies manufacturing germanium diodes and transistors, including IBM, developed this process, which shows the importance of this market.

4.2.2. Electronic silicon production

4.2.2.1. Production of polycrystalline silicon

In the 1950s, given the technological difficulties involved in purifying and crystal-growing silicon, many researchers began to doubt whether it was possible to obtain silicon with satisfactory properties. Nevertheless, under the impetus of the military, for whom temperature stability in transistor operation was an absolute requirement, technological research continued.

The production of polycrystalline silicon involves three main stages:

– production of silicon chlorides from metallurgical silicon produced in an arc furnace (Chapter 2);

– separation of silicon chlorides, metal chlorides and boron chloride;

– polycrystalline silicon production using the CVD process[7]: reduction of chloride ($SiCl_4$) or trichloride ($SiHCl_3$) with hydrogen (Siemens process) or silane decomposition (SiH_4) (Union Carbide process): reaction between gaseous reactants occurring in contact with a silicon-deposited rod, or in contact with purified silicon particles with particle growth.

4.2.2.1.1 Metallic impurities

Silicon is unique when it comes to purification; metallurgical or chemical extraction of impurities is "easy". The various metallic elements have very low solubility in silicon in its solid state. Furthermore, directed solidification purifies silicon by segregating impurities. This is how silicon was purified in its early days.

What is more, most of the metallic impurities in the form of chlorides can be completely extracted from chlorosilanes by distillation.

In addition, the CVD process for polycrystalline silicon production produces additional purification.

Finally, obtaining single crystals by the CZ or FZ process is the most powerful purification step, due to the very low segregation coefficients of these metallic impurities ($k = 6.4 \times 10^{-6}$ for iron); this makes it possible to achieve levels below the requirements of the SIA roadmap (10^8–10^9 atoms/cm^3) (Harrell et al. 1996).

4.2.2.1.2. Boron and phosphorus: basic semiconductor dopants

The same is not true for B and P. To achieve conductivity control, silicon metallurgy must first be purified to a total impurity content of <1 ppba (9N). However, the two elements, boron and phosphorus, introduced into silicon (9N) after purification to obtain the desired conductivity, are present in metallurgical silicon (% P: 20–45 ppmwt; % B: 40–60 ppmwt) (B and P are present in quartz and in reducing agent C) and are the most difficult elements to remove from metallurgical silicon. Hence, the sequence of purification and transformation operations that leads to polycrystalline silicon resistivity >1,000 Ω·cm, 11N purity, corresponds to a concentration of n or p dopants of the order of 5×10^{11}/cm^3 atoms:donors (P, As, Sb) <0.01 ppba and acceptors (B, Al) <0.02 ppba (Chapter 1, Figure 1.13).

---

7 The CVD process was developed by Van Arkel and de Boer in 1925 for the purification of titanium/zirconium by decomposition of titanium iodide (TiI_4) in contact with a tungsten filament.

Phosphorus, boron, arsenic and even sulfur cannot be extracted by simple distillation, even with columns of 100 theoretical plates. Boron and phosphorus chlorides must be transformed into less volatile compounds before distillation by reaction with complexing agents.

Boron chloride BCl_3 can be separated by adsorption on activated silica or activated carbon columns.

With the CVD process used to produce polycrystalline silicon due to the different reaction rates of silicon chloride, boron chloride and phosphorus chloride with hydrogen, the concentration of B and P impurities is also greatly reduced, by factors of up to 1,000.

4.2.2.2. Polycrystalline silicon production processes

During the Second World War, DuPont was the only company to supply 5N (5-9) purity silicon to companies such as ATT, Westinghouse, Sylvania and General Electric, manufacturers of advanced diodes (see Chapter 2).

After the Second World War, Bell Labs developed a process for purifying silicon tetrachloride $SiCl_4$ by reacting silica gel with aluminum chloride $AlCl_3$ (Whelan 1958).

Two industrial processes then came to the fore: the Siemens process and the Union Carbide process. Both processes and their variants derive from processes developed by DuPont de Nemours, namely, the chlorination of metallurgical silicon with the formation of $SiCl_4$ (TET) and $SiHCl_3$ (TCS), which can be purified by distillation, and the decomposition/reduction of the chloride using hydrogen (Siemens process) or by decomposition of SiH_4 (Union Carbide process) on silicon rods.

4.2.2.2.1. Du Pont de Nemours processes

In 1950, Bell Labs asked Union Carbide and DuPont de Nemours to produce high-purity silicon (6N). Although involved in the production of silicones and silicon chloride (see below), Union Carbide was unable to meet the demand. DuPont, on the other hand, took up the challenge, and in 1952, they delivered to Bell Labs 100 kg of "high-purity" silicon produced by the process of reducing $SiCl_4$ with ultrapurified zinc (presented in Chapter 2, section 2.3.1.2) (Lorenz 1984).

The first NPN bipolar transistor in silicon was manufactured by Teal, in May 1954 (formerly Bell Labs) at Texas Instruments (TI), using the "CZ-diffusion" (grown-junction) process (see Chapter 5) with silicon supplied by DuPont at a price of $500 per pound (Riordan and Hoddesson 1997, Chapter 6).

Between 1952 and 1957, DuPont developed and patented a number of processes for purifying silicon chloride and zinc to the required purities (Olson 1988).

However, the DuPont process (SiCl$_4$ + Zn) had one drawback: it was unable to eliminate boron. The boron chloride (BCl$_3$) present was reduced to boron, simultaneously with SiCl$_4$, by the zinc and incorporated (absorbed) into the silicon produced.

In 1957, DuPont developed the CVD process for reducing silicon chloride SiCl$_4$ or trichloride SiHCl$_3$ with hydrogen on silicon particles suspended in a fluidized bed at 1,040°C. The boron content of the particles produced was <2 ppb and the phosphorus content <3 ppb. Hydrogen, being a weaker reducing agent than zinc, did not reduce boron chloride, which could be removed by distillation. For this reason, the Siemens process uses hydrogen as a reducing agent for silicon trichloride. This fluidized-bed reduction process was taken over by Ethyl Corporation Process (based on a TI patent).

With the invention of the planar bipolar transistor and integrated circuits, it became imperative to produce silicon with a concentration of impurities down to the ppb level from the ppm range. To meet this demand, DuPont developed a CVD process for producing silicon by thermal decomposition of SiH$_4$ silane on the surface of an ultrapure silicon rod (a process later adopted and developed by Union Carbide). Until 1961, DuPont continued to be the main supplier of high-purity silicon. It withdrew from this market in 1962, having no other use for the product (Bertrand and Colson 1961).

To consolidate its leadership in the production of silicon transistors, TI began in 1954, under the impetus of Gordon Teal, to develop the high-purity silicon production process (Teal 1976); this produced high-purity silicon tetrachloride and reduced it with hydrogen.

In December 1956, TI became a major supplier of high-purity silicon to the semiconductor industry. By the mid-1960s, TI was producing half of the total electronic silicon and was one of only three manufacturers of silicon single crystals.

4.2.2.2.2. Siemens process

Siemens recognized the importance of semiconductors very soon after the war, with Walter Schottky on its team of experts[8]. Silicon point-contact diodes had been developed in Germany during the war.

---

8 Walter Schottky in Germany and Nevill F. Mott in England had independently explained the rectifying effect of the metal-semiconductor junction just before the war (see Chapter 2).

The Siemens-Schuckert division, with Schottky and Spenke as experts, worked on high-power rectifiers and switches; the Siemens-Halske division worked on telephony studying germanium components from 1945.

The Siemens-Schuckert division set up a laboratory in Bavaria in 1946, headed by Spenke, to develop electronic silicon (Spenke and Heywang 1981).

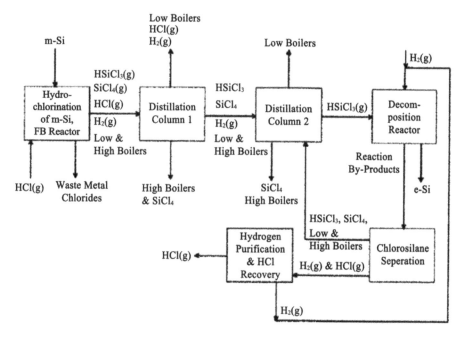

Figure 4.11. *Siemens process for producing polycrystalline electronic silicon (Jiao et al. 2011)*

The Siemens process (Figure 4.11) comprises three stages: the first is the production of trichlorosilane from silicon metal, the second is the purification of trichlorosilane by distillation, and the third is hydrogen reduction of trichlorosilane (TCS) by the CVD process on inductively heated silicon rods (Gutsche 1962) (Figure 4.12):

$$SiHCl_3 + H_2 \rightarrow Si(s) + 3\ HCl$$

Trichlorosilane was chosen for its deposition speed and low boiling point (31.8°C). The boiling temperatures of the other chlorides allow their separation by fractional distillation ($SiCl_4$: 57.6°C; SiH_2Cl_2: 8.6°C). In this final operation for the production of polycrystalline silicon, the concentration of impurities is reduced – in the case of boron and phosphorus, by a factor of 1,000.

The silicon rods are then purified using the ZM process to produce single crystals. A resistivity of 500 Ω·cm was reached in 1954 and 5,000 Ω·cm in 1957 (see Chapter 1, Figure 1.13).

Siemens granted a license to manufacture silicon electronics to Wacker in 1958, a chemical company that had been manufacturing silicones since the 1930s.

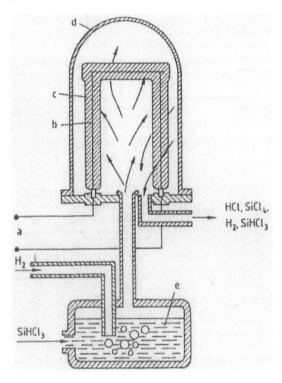

Figure 4.12. *Siemens CVD process for producing silicon rods ($SiHCl_3$ + H_2) (Gutsche 1962)*

4.2.2.2.3. Union Carbide process

Union Carbide had been involved in silicon chemistry since the 1930s and again after the war, conducting research into silicones and numerous silicon compounds, including chlorides. In the 1950s, trichloro-silane with extremely low levels of B and P was obtained.

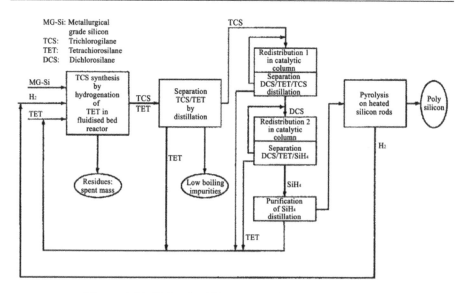

Figure 4.13. *Union Carbide process for the production of polycrystalline electronic silicon (Jiao et al. 2011)*

TI, which, since 1954 (see above), had been producing high-purity silicon by reducing silicon tetrachloride ($SiCl_4$) with hydrogen, asked Union Carbide to supply it with the purest silicon chloride it could produce. Union Carbide set about industrially producing silicon trichloride, while developing purification methods to achieve B and P contents of less than 0.05 ppb.

In the 1960s, Union Carbide began to produce silane industrially.

With the energy crisis of 1974–1975, at the instigation of the American government, Union Carbide joined the Low Cost Solar Array Project and set about developing a process for producing silicon from silane, believing it to be less costly than the Siemens process.

The Union Carbide process (Figure 4.13) (Lorenz 1984; Breneman 1987; Jiao et al. 2011) comprises three main stages:

– metallurgical silicon conversion in a fluidized bed reactor into silicon trichloride $SiHCl_3$ (TCS) by hydrogenation of $SiCl_4$ (TET) recycled by the reaction ($Si + SiCl_4 + H_2 \rightarrow SiHCl_3$);

– catalytic conversion of TCS to silane SiH_4;

– pyrolysis (decomposition) of silane on silicon rods ($SiH_4 \rightarrow Si + 2H_2$). The process uses the same type of reactor as the Siemens reactor.

In 1984, this process, originally developed for the solar industry, was adopted by the electronics industry.

4.2.3. *Single crystal production*

4.2.3.1. *CZ process*

During his last two years (1951–1952) at Bell Labs, Gordon Teal, with the help of Ernie Buehler, succeeded in manufacturing silicon single crystals using the CZ process (Figure 4.14) by modifying the equipment developed for the production of germanium monocrystals: pulling under a controlled atmosphere to avoid silicon oxidation and replacement of the graphite crucible by a quartz crucible (Figure 4.15) (Teal and Buehler 1952; Teal 1976).

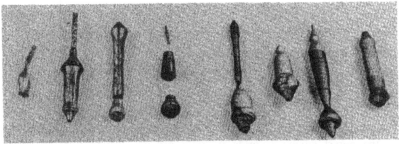

Fig. 24. Early single crystals of silicon grown by G.K. Teal and E. Buehler

Fig. 25. Single crystal of silicon #136. Length: 9-10 in.

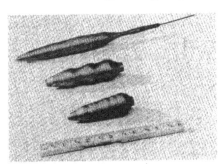

Fig 27. Highly symmetrial silicon single crystal. Length: $4^{1/4}$ in.

Fig 26. Three silicon single crystals. Lengths: $4^{1/2}$, 3 and $2^{3/4}$ in.

Figure 4.14. *First silicon single crystals manufactured by Teal and Buehler (Teal 1976)*

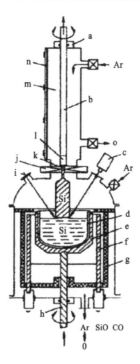

Figure 4.15. *Apparatus for pulling a silicon single crystal using the CZ (Czochralski) process (speed 1 mm/min for Φ = 150 mm) (Glisenti 1962)*

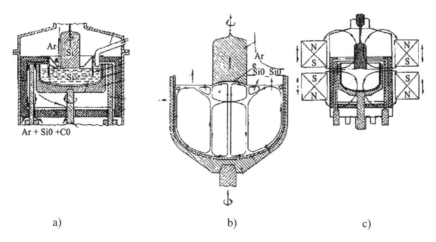

a)   b)   c)

Figure 4.16. *Evolution of the CZ process: (a) continuous feed; (b) convection movements; (c) vertical magnetic fields to counteract convection movements (b) (Glisenti 1962)*

However, as a result of the dissolution of the silica in the crucible, enhanced by the convective motions of the liquid silicon bath, oxygen is incorporated into the silicon and present in concentrations of the order of 20–40 ppma ($1–2 \times 10^{18}$ a/cm^3).

This problem has prompted numerous studies aimed at reducing these convection movements (Figure 4.16(b)). Hence, the use of magnetic fields perpendicular to the crystal growth axis to reduce convection currents (Figure 4.16(c)). Oxygen content is then reduced to values $<10^{18}$ atoms/cm^3 (with the ZM process, oxygen content is of the order of 1×10^{16} atoms/cm)^3.

Carbon is present in concentrations from 10^{16} up to 3×10^{17} atoms/cm^3 (i.e. 6,000 ppba or 2,500 ppbwt), the solubility limit. It comes from the graphite components of the crystal pulling plant.

At present, the crucibles used in the CZ process are made from quartz sand with a purity of 99.998% Si (Al < 16 ppm; Ti < 1 ppm).

4.2.3.2. *Production of single-crystal silicon without structural defects (dislocations)*

Screw dislocations are traps for minority carriers, and impurities, especially heavy metals, tend to precipitate (segregate) on dislocations, as does oxygen.

Following observations concerning the detrimental influence of dislocations on the performance of diodes and transistors, modifications to the production equipment and to the silicon single-crystal pulling procedure made it possible (in the 1960s) to obtain crystals with an almost perfect structure (absence of dislocations) using the Dash process, developed by William Dash of General Electric (Dash 1959). When the seed crystal first comes into contact with the liquid bath, a high pulling speed produces a crystal a few millimeters in diameter, and the dislocations present in the seed crystal do not develop in the growing crystal. The pulling speed is then reduced and the diameter of the growing crystal increases.

Most materials crystallize, forming defects and imperfections. Silicon is unique in that it forms perfect (virtually defect-free) crystals. These can be obtained using either the CZ process or the FZ process.

4.2.3.3. *The FZ process*

The FZ (float-zone) process for manufacturing single crystals, derived from the ZM process, was developed by Henry Theuerer in 1956 (Figure 4.17). In this process, the crucible is eliminated. A molten zone is formed locally in a vertical bar. It is stabilized by the surface tension of the liquid. The molten zone is displaced along the bar. The process enables both purification and single-crystal growth in a single operation, starting from a seed of given orientation. It produces silicon of comparable purity to germanium. However, it does not produce rods larger than 10 cm in diameter. This process was eclipsed by the CZ process, which produces rods up to 300 and even 450 mm in diameter (Theuerer 1962).

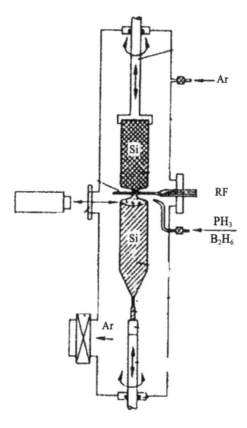

Figure 4.17. *Apparatus for growing a silicon single crystal using the FZ process (speed 2 mm/min for Φ = 150 mm). Upper part Si poly-crystal, lower part Si single crystal. Doping (PH3 or BH3) is carried out during growth (Glisenti 1962)*

4.2.4. *Manufacture of a silicon diode using the CZ process*

In 1952, although the characteristics of the first single crystals were mediocre due to the presence of relatively high levels of oxygen, the first silicon PN diode was manufactured by Pearson using the CZ process utilized to produce the first germanium diodes by Pearson and Sawyer of Bell Labs.

This diode proved superior to germanium diodes, with a leakage current 1,000 times lower, and could operate up to 300°C (Pearson and Sawyer 1952). "Immediately silicon became a rival to germanium" (Leamy and Wernick 1997).

4.2.5. Industrial developments in single-crystal pulling processes

In 1954, Gordon Teal, having left Bell Labs and joined TI, after developing the production of "electronic" silicon, then set about developing the manufacture of single crystals using the CZ process, which enabled TI to dominate silicon transistor production until 1958.

At the same time, Fairchild Semiconductor began producing silicon single crystals using the ZM process by the team that had left Shockley Semiconductor Laboratory Semiconductor Laboratory in 1957 and formed Fairchild. Development at Shockley Semiconductor had been a major fiasco.

At the same time, a Shockley Semiconductor engineer, Dean Knapic, set up his own company, Knapic Electrophysics, to produce silicon single crystals. He soon developed the ZM process. However, the company lost money in the first few years and the financiers closed it down.

One of his engineers, Lorenzini (2007), bought some of Knapic's equipment and set up his own company, Elmat Corporation, which soon had customers including RCA and even TI. Elmat was eventually acquired by General Transistor/General Instruments (supplier of transistors for Cray's CDC 1604 computers).

Lorenzini founded the Siltec company in 1969, which became one of the two companies producing silicon wafers, the base wafers for transistors obtained from single crystals. Siltec was acquired by Mitsubishi Metal in 1986.

In 1959, Monsanto founded MEMC (Monsanto Electronic Materials Company, a silicon wafer manufacturing division), which later became Sun Edison, the last American company to manufacture silicon wafers. In February 1960, MEMC began production of single crystals using the CZ process with a diameter of 19 mm; and in 1966, began production of structurally flawless crystals (a process developed by William Dash in 1959).

In 1979, MEMC became the first company to manufacture single crystals with a diameter of 125 mm (5 inches); and in 1984, in partnership with IBM, MEMC produced slices with diameters of 200 mm (Figure 4.18) and 300 mm (Figure 4.19) (for a history of the MEMC, see referenceforbusiness.com).

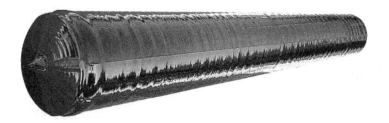

Figure 4.18. *Silicon single crystal, diameter 200 mm (150 kg). Photo: GZV Optics. For a color version of this figure, see www.iste.co.uk/vignes/silicon1.zip*

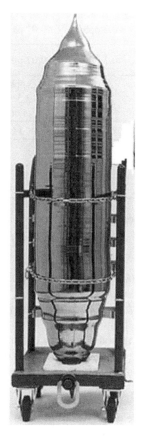

Figure 4.19. *300 mm diameter single crystal (Courtesy of MEMC). For a color version of this figure, see www.iste.co.uk/vignes/silicon1.zip*

Table 4.1 shows the chemical and physical characteristics of silicon at different stages of production and purification.

Element/ characteristic	Silicon metallurgical	Trichlorosilane (TCS)	CVD- Si/EG-Si	CZ-Si	FZ-Si
C		<0.3 ppma	<200 ppba	<500 ppba 2.5 × 10^{16} a/cm^3	
O		–	<100	20–40 ppma/ 1–2 × 10^{18} a/cm^3	<20 ppba 1 × 10^{16} a/cm^3
Dopant type p (B)	44 ± 13 ppma	<0.07 ppba	<0.02 ppba		
Dopant type n (P+AS)	28 ± 6 ppma	<0.2 ppba	<0.01 ppba		
Al	1,570 ± 580 ppma	<0.3 ppba	<0.01 ppba		
Fe	2,070 ± 510 ppma	<2 ppbw (5 ppba)	<0.5 ppba → 2.5 × 10^{11} a/cm^3	0.005 ppba < 10^8– 10^9 a/cm^3	
Resistivity				2,000 Ω	>2,000 Ω·cm	5,000 Ω·cm
Lifetime of minority charge carriers				30–300 μs	50–500 μs

Table 4.1. *Chemical and physical characteristics of silicon (sources: TCS: Air Products; CZ-Si and FZ-SI: Zulehner (2003))*

4.3. Appendix: physical basis of PN diode operation

4.3.1. *Energy band diagram and potential barrier*

When an N region and a P region come into contact, due to the difference in free charge carrier concentrations between the two regions, or Fermi levels (Figure 4.20) (see Chapter 1, section 1.2.4 and Figure 1.7), electrons from the N region cross the junction and "pass" into the P region, and vice versa for holes from the P region.

However, in doing so, they leave a layer of positive ions from the "donor" atoms in the N region near the junction, which then takes on the charge +. Similarly, a layer of negative ions is formed in the P region, which takes on the charge –. A depletion layer is thus formed on either side of the junction forming a *potential barrier*. The curvature of the conduction band boundary in the depletion layer reflects the variation in the density of free charge carriers on either side of the interface between the regions.

This potential barrier, which increases as electrons move from the N region to the P region and as holes move from the P region to the N region, opposes their further passage. Migration stops when the Fermi levels are aligned. The potential barrier reaches V_{bi} (built-in potential, "threshold voltage "), as shown in Figure 4.21:

$$eV_{bi} = E_{Cp} - E_{Cn} \qquad [4.1a]$$

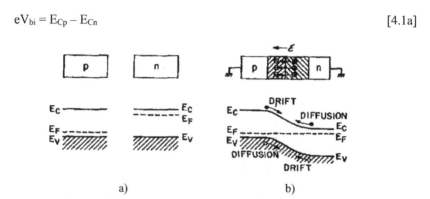

Figure 4.20. *PN diode: energy band diagram: (a) before contact; (b) after contact between regions. Formation of a depletion layer forming a potential barrier V_{bi} (Sze 2002, p. 88)*

The potential barrier is a function of the respective doping of the two regions: N_D donor concentration and N_A acceptor concentration[9]:

$$V_{bi} = (kT/e) \ln (N_D N_A/n_i^2) \qquad [4.1b]$$

At equilibrium, the concentration of electrons in the P region, $n_{(P)0}$, and that of holes in the N region, $p_{(N)0}$, are functions of the potential V_{bi} (Figure 4.22):

$$n_{(P)0} = n_{(N)0}. \exp (-eV_{bi}/kT) \text{ and } p_{(N)0} = p_{(P)0}. \exp (-eV_{bi}/kT) \qquad [4.2]$$

where $n_{(N)0} = N_D$ is the concentration of electrons in the N region and $p_{(P)0} = N_A$ the concentration of holes in the P region.

They satisfy the law of mass action (Chapter 1, formula [1.2]):

$$n_{(P)0}.p_{(P)0} = n_{(P)0} (N_A) = n_i^2$$

---

9 All expressions (formulas) given in this chapter are taken from Sze (2002, p. 88).

$$p_{(N)0}.n_{(N)0} = p_{(N)0} (N_D) = n_i^2 \tag{4.3}$$

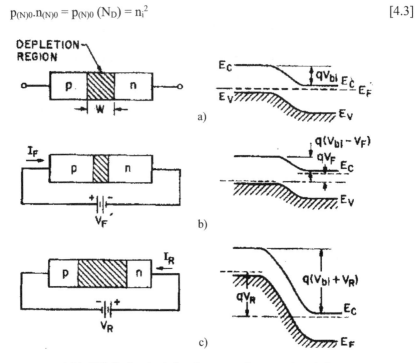

Figure 4.21. *PN diode: depletion layers and energy band diagrams.*
(a) Equilibrium junction and potential barrier V_{bi}; (b) forward-biased
junction; (c) reverse-biased junction (Sze 2010, p. 97)

In forward bias, the potential barrier is lowered and becomes $(V_{bi} - V_F)$ (Figure 4.21(b)), allowing electrons from the N region to pass into the P region and holes from the P region to pass into the N region.

For an applied voltage V_F, the respective concentrations of electrons $n_{(P)}$ and holes $p_{(N)}$ (excess carrier) on either side of this barrier layer are (Figures 4.4(a) and 4.22(a)) (see formula [4.2]):

$$n_{(P)}(x_p) = n_{(N)0}. \exp \{e(V_F-V_{bi})/kT\} \text{ and } p_{(N)}(x_N) = p_{(P)0}. \exp \{e(V_F-V_{bi})/kT\}$$

Or, using formula [4.3]:

$$n_{(P)}(x_p) = \{n^2_i/N_A\} \exp(eV_F/kT) > n_{(P)0} \tag{4.4}$$

$$p_{(N)}(x_N) = \{n^2_i/N_D\} \exp(eV_F/kT) > p_{(N)0} \tag{4.5}$$

4.3.2. *Electron and hole currents: "lifetime" and "diffusion length" of "minority carriers"*

The difference in electron concentration in the P region, $(n_{(P)}(x_p) - n_{(P)0})$, and that of holes in the N region $(p_{(N)}(x_n) - p_{n0})$, induces an electron current in the P region and a hole current in the N region, as a function of the respective concentration gradients given by the first law of diffusion (Fick's law):

$$I_n(-x_p) = -D_{n(P)} \, (dn_p/dx) \text{ and } I_p(-x_n) = -D_{p(N)} \, (dp_n/dx) \qquad [4.6]$$

where $D_{n(P)}$ is the diffusion coefficient of electrons (minority carriers) in the P region and $D_{p(N)}$ the diffusion coefficient of holes in the N region.

As they move, the minority carriers (electrons in the P region and holes in the N region) gradually recombine with the majority carriers in the corresponding region on the electron–hole recombination centers (traps) (Figure 4.3) (Chapter 1, section 1.2.5 and Figure 1.7).

In the P region, the electron–hole recombination speed for an electron is expressed by (Chapter 1, section 1.2.5.2, formula [1.9a]):

$$r = \Delta n/\tau_\pi = (n_{(P)} - n_{(P)0})/\tau_{Rn(P)} \qquad [4.7a]$$

where $\tau_{n(P)}$ is the "lifetime" of the electron in the P region (minority carrier lifetime). The same applies to a hole coming from the P region and injected into the N region: the recombination velocity is expressed by:

$$r = \Delta p/\tau_{\pi(N)} = r = (p_{(N)} - p_{(N)0})/\tau_{Rp(N)} \qquad [4.7b]$$

The lifetimes are equal: $\tau_n(P) = \tau_p(N) = \tau_m$, the concentration of electron–hole recombination centers (traps) being the same throughout the single crystal making up the diode, and hole capture and emission coefficients being equal.

Since recombination is progressive, hole and electron currents are controlled by electron–hole recombination, electrons in the P region and holes in the N region in forward and reverse polarization (Figure 4.23).

The hole and electron currents are gained by solving the second law of diffusion, which, in steady state, is written as:

$$Dp_{(N)} \cdot d^2p_{(n)}/dx^2 = (p_{(N)} - p_{(n)0})/tp \qquad [4.8]$$

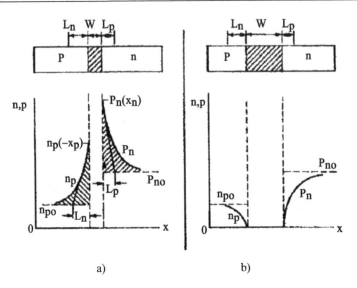

Figure 4.22. *Concentration profiles of minority carriers in each region:*
(a) under direct polarization; (b) under reverse polarity (Sze 2002, p. 107)

The current of electrons leaving the N region and passing into the P region and the hole current leaving the P region and passing into the N region are given by:

$$I_{n(xp)} = e n_i^2 (D_{n(P)}/N_A L_{n(P)}) \cdot \{\exp(eV/kT) - 1\}$$ [4.9a]

$$I_p(x_n) = e n_i^2 (D_{p(N)}/N_D L_{p(N)}) \cdot \{\exp(eV/kT) - 1\}$$ [4.9b]

Two parameters appear, $L_{n(P)}$ and $L_{p(N)}$, each as a function of the "lifetime" of the minority carriers in each region (these parameters can be deduced directly from formula [4.8]):

$$L_{n(P)} = (D_{n(P)} \cdot \tau_{n(P)})^{1/2} \text{ and } L_{p(N)} = (D_{p(N)} \cdot \tau_{p(N)})^{1/2}$$ [4.10]

Having the dimension of a length called the "diffusion length", this represents the average distance traveled by electrons in the P region before they are recombined by holes in the N region (Figure 4.22).

The diffusion coefficient $D_{n(P)}$ is related to the mobility of the charge carrier μ_n (defined in Chapter 1, section 1.2.8) by Einstein's relation:

$$D_{n(P)} = (kT/e) \cdot \mu_n$$ [4.11]

4.3.3. *Current in the diode*

The total current I (per unit area), in positive polarity, flowing through the diode is the sum of the diffusion currents in each phase (Figure 4.22):

$$I = I_{n(P)(xp)} + I_{p(N(xn))} \qquad\qquad [4.12]$$

or:

$$I = en_i^2 \, (D_{n(P)}/N_A L_{n(P)} + D_{p(N)}/N_D L_{p(N)}) \cdot \{\exp(eV/kT) - 1\}$$

$$I_S \cdot \{\exp(eV/kT) - 1\} \qquad\qquad [4.13]$$

The ratio of the electron current I_n to the hole current I_p is proportional to the ratio of the concentration of n-type dopants (N_D) to the concentration of p-type dopants (N_A) respectively in each region, the lifetimes being equal:

$$I_n/I_p = D_{n(P)}/D_{p(N)} \cdot (L_{p(N)}/L_{n(P)}) \cdot \{N_D/N_A\} \approx (\mu_n/\mu_P)^{1/2} \cdot \{N_D/N_A\} \qquad [4.14]$$

The diffusion length is the quantity we aim to maximize (in the case of solar cells and bipolar amplifier transistors) or minimize (in the case of diodes and transistors). It is controlled by the phenomenon of electron–hole recombination, itself controlled by the "impurities" present (section 4.1.5).

In reverse polarization, the total current is the sum of an "electron current" and a "hole current". They are therefore given by the same expressions [4.9], the potential V_R being negative, the exponential tends toward 1 very quickly. The sum of these currents is the saturation current I_S:

$$I_S = en_i^2 \, (D_{n(P)}/N_A L_{n(P)} + D_{p(N)}/N_D L_{p(N)}) \qquad\qquad [4.15]$$

A second source of current is due to the generation of electron–hole pairs in the depleted layer. In the depletion layer, the rate of generation, in inverted polarity, is proportional to the equilibrium concentration of intrinsic carriers (formulas [1.1] and [4.7a]):

$$r_G = n_i/2\tau_m \qquad\qquad [4.16]$$

and the induced current has the expression:

$$I_{gen} = en_i \, W/2\tau_m \qquad\qquad [4.17]$$

where W is the depletion layer thickness that depends on the (inverse) potential applied to the junction.

4.4. References

Bertrand, L. and Colson, C.M. (1961). Silicon production. Patent, US3012862A.

Breneman, W.C. (1987). High purity silane and silicon production. Patent, US4676967.

Ciccolalla, D.F. (1962). Semiconductor diodes as fast operating switches. Patent, US3067485.

Collins, C.B. (1957). Properties of gold–doped silicon. *Physical Review*, 105(4), 1168–1173.

Dash, W.C. (1958a). Evidence of dislocation jogs in deformed silicon. *Journal of Applied Physics*, 29, 705.

Dash, W.C. (1958b). Silicon crystals free of dislocations. *Journal of Applied Physics*, 29, 736.

Dash, W.C. (1959). Growth of silicon crystals free from dislocations. *Journal of Applied Physics*, 30, 459.

Glisenti, A. (1962). Photovoltaics: Si-based solar cells. University of Padua [Online]. Available at: https://www.dei.unipd.it/en/electronics/photovoltaic-devices.

Gutsche, H. (1962). Method for producing highest purity silicon for electric semiconductor devices. Patent, US3042494.

Harrell, S., Seidel, T., Fay, B. (1996). The national technology roadmap for semiconductors. *Microelectronic Engineering*, 30, 1(4), 11–15.

Haynes, J.R. and Shockley, W. (1951). The mobility and life of injected holes and electrons in germanium. *Physical Review*, 81, 835.

Hoyem, A.G. and Tyndall, E.P.T. (1929). An experimental study of the growth of zinc crystals by the CZ method. *Physical Review*, 33, 81.

Hu, C. (2009). *Modern Semiconductor Devices for Integrated Circuits*. Pearson, London.

Huff, H.R. (2002). An electronics division retrospective (1952-2002) and future opportunities in the twenty-first century. *Journal of the Electrochemical Society*, 149(5), 35–58.

Istratov, A.A., Buonassisi, T., Pickett, M.D., Heuer, M., Weber, E.R. (2006). Control of metal impurities in "dirty" multicrystalline silicon for solar cells. *Materials Science and Engineering*, 134, 282–286.

Jiao, Y., Salce, A., Ben, W., Jiao, Y., Salce, A., Ben, W., Jiang, F., JI, X., Morey, E., Lynch, D. et al. (2011). Siemens and Siemens-like processes. *JOM*, 63(1), 28.

Koeniguer, C. (n.d.). *Les diodes* [Online]. Available at: cedric.koeniguer.fres.fr/documents/cours/chapitre1/les diodes.

Kyropoulos, S. (1926). Ein Verfahren zur Herstellung grosser Kristalle. *Zeitschrift für anorganische und allgemeine Chemie*, 154, 308–313.

Leamy, H.J. and Wernick, J.H. (1997). Semiconductor silicon: The extraordinary made ordinary. *MRS Bulletin*, 5, 47–55.

Lecuyer, C. and Brock, D.C. (2006). The materiality of microelectronics. *History and Technology*, 22(3), 301–325.

Little, J.B. and Teal, G.K. (1954). Methods of producing semiconductive bodies. Patent, US2683676.

Lorenz, J.H. (1984). *The Silicon Challenge: Silicon Material Preparation and Economical Wafering Methods*. Noyes Publication, Norwich.

Lorenzini, R.E. (2007). Interview "Science Industry Institute" [Online]. Available at: https://digital.sciencehistory.org.

Meroli, S. (2012). The minority carrier lifetime in silicon wafer [Online]. Available at: meroli.web.cern.ch/meroli/lecture_lifetime.

Monsanto Electronic Materials Company (MEMC) [Online]. Available at: https://www.referenceforbusiness.

Newman, R.C. (1982). Defects in silicon. The Institute of Physics. *Reports on Progress in Physics*, 45, 1163.

Ohl, R.S. (1946a). Light-sensitive electric device. Patent, 2402662.

Ohl, R.S. (1946b). Alternating current rectifier. Patent, US2402661.

Olson, C.M. (1988). The pure stuff. *American Heritage of Invention and Technology*, 4(1), 58.

Pearson, G.L. and Brattain, W. (1955). History of semiconductor research. *Proceedings of the IRE*, 12, 1794–1806.

Pearson, G.I. and Sawyer, S.B. (1952). Silicon P-N-junction alloy diodes I.E.E.E. *Proceedings of the IRE*, 40(11), 1348–1351.

Pfann, W.G. (1952). Principles of zone melting. *Transactions of the American Institute of Mining and Metallurgical Engineers*, 194, 747–753.

Pfann, W.G. (1956). Process for controlling solute segregation by zone melting. Patent, US2739088.

Pfann, W.G. (1966). *Zone Melting*, 2nd edition. Wiley, New York.

Riordan, M. and Hoddeson, L. (1997). *Crystal Fire: The Invention of the Transistor and the Birth of the Information Age*. W.W. Norton & Company, New York.

Riordan, M., Hoddesson, L., Herring, C. (1999). The invention of the transistor. *Review of Modern Physics*, 71(2), 336–345.

Saby, J.E. (1952). Fused impurity P-N-P junction transistors. *Proceedings of the IRE*, 40, 1358–1360.

Scaff, J.H. (1946). Preparation of silicon materials. Patent, US2402582A.

Scaff, J.H. (1970). The role of metallurgy in the technology of electronic materials. *Metallurgical Transactions*, 1(3), 561–572.

Seitz, F. and Einspruch, N.G. (1998). *Electronic Genie: The Tangled History of Silicon*. University of Illinois Press, Illinois.

Shockley, W., Sparks, M., Teal, G.K. (1951). P-n junction, transistors. *Physical Review*, 83, 151–162.

Sopori, B. (1999). Impurities and defects in photovoltaic devices. A review. In *Conference: 10th International Workshop on the Physics of Semiconductor Devices*. NREL (National Renewable Energy Laboratory), Delhi.

Spenke, E. and Heywang, W. (1981). Twenty five years of semiconductor-grade silicon. *Physica Statu Solidi*, 64, 11.

Storks, K.H. and Teal, G.K. (1948). Electrical translating materials and method of making them. Patent, US2441603.

Sze, S.M. (2002). *Semiconductor Devices*, 2nd edition. Wiley, New York.

Teal, G.K. (1955). Method of producing semiconductive bodies. Patent, US2727840.

Teal, G.K. (1976). Single crystals of germanium and silicon. *IEEE Transactions on Electron Devices*, 23(7), 621–639.

Teal, G.K. and Buehler, E. (1952). Growth of silicon single crystals and of single crystal silicon p-n junctions. *Physical Review*, 87, 190.

Theuerer, H.C. (1962). Method of producing semiconductive materials. Patent, US3060123.

Weber, E.R. (1983). Transition metals in silicon. *Applied Physics*, 30, 1–22.

Whelan, J.M. (1958). Method of purifying silicon tetrachloride and germanium tetrachloride. Patent, US2821460.

Zulehner, W. (2003). *Ullmann's Encyclopedia of Industrial Chemistry*, volume 32. Wiley-VCH, Weinheim.

5

The Bipolar Transistor

The invention of the bipolar transistor by William Shockley in 1949 and the invention of the transistor's "planar" structure by Jean Hoerni in 1959 ushered in the digital revolution era.

The concept of this transistor follows from the discovery of the PN junction by Ohl in 1940 and the mechanism formulated by Bardeen and Brattain in 1948 to explain the functioning of the point-contact transistor.

The concept was validated by the 1951 production of the first transistor, based on germanium, using the CZ single-crystal pulling process.

The realization of this transistor led to the development of processes enabling the realization of increasingly high-performance transistors, which in turn led to the development of commercial computers equipped with such transistors.

The first bipolar silicon transistors appeared in 1954, and those with a planar structure in 1959.

Germanium components (diodes, bipolar transistors) dominated the logic circuits of "calculators" in the 1950s (including IBM 608, until 1965).

Bipolar silicon transistors were used in computers (including the IBM 360) from 1964 onwards.

Germanium and silicon bipolar transistors equipped the Vanguard 1 and Explorer 1 and 3 satellites in 1958.

While bipolar transistors have been replaced by MOSFET transistors in digital applications for their switching function, for their function of amplifying weak signals, they are the basic component of telecommunications amplification circuits.

This chapter presents:

– the invention of the bipolar transistor;

– the transistor operation (in the static regime);

– the transistor functions;

– the manufacturing technologies;

– the "mesa" and "planar" silicon bipolar transistors;

– the industrial development of bipolar transistors.

5.1. Transistor operation and functions

A. Fleming invented the vacuum diode in 1903, and in 1906, Lee de Forest invented the triode by adding a grid between the diode's cathode and anode. In addition to rectifying the alternating current, a small signal variation on the grid resulted in an amplification of the cathode–anode current. A change in the signal applied to the gate switches the triode on or off, and it can also act as a switch. The triode is also capable of self-oscillation, hence its use in radio transmitters and receivers.

The bipolar transistor is a component that can perform the same functions – amplification of weak currents and switching – much more quickly.

The operation of the triode, like transistors, relies on the control of a current of electrons, which can either be amplified, interrupted or reignited.

In the triode, however, switching times are much longer and the permissible frequencies much lower than in solid-state components, as they are related to the time taken by electrons to cross the distance (approximately 1 mm) between cathode and anode (transit time); whereas in a bipolar transistor, the distance covered by electrons between emitter and collector is less than 1 μm, down to around 10 nm.

5.1.1. *History*

5.1.1.1. *Shockley's invention*

The concept of the bipolar junction transistor (BJT) was formulated by William Shockley on January 23, 1948 (remember that it was in December 1947 that

Bardeen and Brattain discovered the point-contact transistor). He was awarded the Nobel Prize for this invention[1].

The discovery of the germanium N point-contact transistor, in December 1947, encouraged Shockley to search for a component stronger without the contact points. This led almost directly to the design of what was to become the NPN bipolar transistor in January 1948 (Figure 5.1); this was an invention he called "semiconductor valve; a valve like the grid in a vacuum tube, it provided a convenient handle to regulate the electron flow through the device" (Riordan and Hoddesson 1997, p. 148).

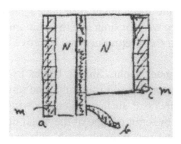

Figure 5.1. *Shockley's sketch of the NPN transistor concept (Shockley 1948, p. 130)*

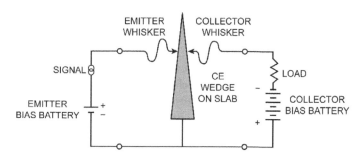

Figure 5.2. *Circuit designed by Shive (1949)*

It was the result of an experiment carried out by John Shive as part of his ongoing work on the point-contact transistor, which led Shockley to think that the device imagined to improve the point-contact transistor was a transistor: "A new way of making a semiconductor amplifier was staring me in the face" (Shockley 1976; Burgess 2008a).

---

1 For a complete history, see Shockley (1976).

Shockley patented the "bipolar transistor" on June 26, 1948 (Shockley 1951) and published a paper in July 1949 detailing the operation of the bipolar transistor (Shockley 1949).

However, it was not until two years later, in 1951, that Bell Labs scientists and engineers developed processes for the industrial manufacture of such transistors (described in section 5.2).

5.1.1.2. *Shive's experiment and the "lifetime" of minority shareholders*

Following the discovery of the "point-contact transistor" in December 1947 by Bardeen and Brattain and the mechanism formulated for its operation (Chapter 3, section 3.2.2), this discovery was considered extremely important, and an entire Bell Labs team was engaged in tests to reproduce and improve the component.

Therefore, a team member (John Shive) began by trying to reproduce the Bardeen and Brattain transistor. Having found that the result was well obtained with a germanium N wafer, with no P surface layer, showing that the positive charge carriers – holes injected by the emitter (one of the metal tips) into the base – reached the collector (the other metal tip) through the N substrate. On February 16, 1948, Shive produced a component in which the transmitter and receiver electrodes were placed on either side of the germanium N wafer (Figure 5.2). With this component, he reproduced Bardeen and Brattain's experiment (Figure 3.9). Shive thus demonstrated that charge carriers of one type (minority carriers) could move in a substrate doped with charge carriers of the opposite sign (majority carriers) (Shive 1949).

The concept of an NPN bipolar transistor based on electron injection from the N emitter into the P base and the transfer of electrons through this base to the collector, and conversely for a PNP bipolar transistor with hole injection, was validated by the performance of germanium single crystals manufactured by Gordon Teal.

Following Shive's demonstration of the injection of minority carriers into a substrate of oppositely charged carriers and their displacement without recombination with the majority carriers , at Shockley's request, the measurement of the lifetime of these minority carriers (Chapter 1, section 1.2.5 and Chapter 4, section 4.3.1) by Haynes in 1948 was the crucial experiment, proving that carriers injected into a region of different polarity could have a sufficient lifetime to validate the bipolar transistor concept. Haynes' first tests were carried out on a thin slice of germanium extracted from a germanium N polycrystal, and the lifetime of the injected charge carriers (holes) was of the order of 10 μs (Haines and Shockley 1949).

Teal obtained germanium single crystals in late 1948 (of little interest to Shockley) (Teal 1976); in early 1949, Haynes repeated measurements "of minority carrier lifetimes on a germanium N single-crystal and obtained lifetimes in excess of 140 microseconds, well above those measured on polycrystals" (Shockley et al. 1949). It was in the light of these results that Shockley completely changed his mind and admitted the major interest of the single crystal for the development of the transistor (see Chapter 4, section 4.2.1.2). In 1951, on Teal's germanium N single crystals with a high degree of perfection, the measured lifetime was in excess of 200 μs (Haynes and Shockley 1951).

5.1.2. *Bipolar transistor operation*

5.1.2.1. *Operating modes*

The NPN bipolar transistor consists of two N regions separated by a P region, that is, two diodes. The characteristic curve, current I_C (flowing through the transistor), as a function of the potential V_{CE} applied to the collector, with, as parameter, the potential applied to the base P V_{BE}, shows two regions corresponding to different operating regimes (modes).

In saturated operation, as long as $V_{CE} < V_{BE}$, current I_C increases with applied voltage V_{CE}, and in active operation $V_{CE} > V_{BE}$, the current I_C is independent of the voltage V_{CE} (Figures 5.3 and 5.4(a)) and dependent on the potential applied to the base V_{BE} (Figure 5.4(b)). The dashed curve delimits these two operating regions.

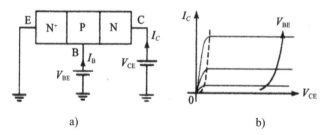

a)   b)

Figure 5.3. *The NPN transistor: (a) "common emitter" configuration; (b) characteristic curves I_C -V_{CE}, with parameter V_{BE} (Hu 2009, p. 293)*

These characteristic curves ($I_C - V_{CE}$) are often drawn with parameter: current I_B (Figure 5.4(a)), since currents I_B and I_C are independent, as will be seen later. This representation enables amplification to be visualized (Figure 5.9).

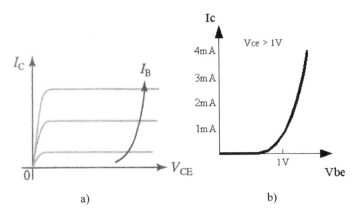

a)   b)

Figure 5.4. *Characteristic curves: (a) Ic vs. VCE, with parameter base current IB;*
(b) Ic vs. VBE in active mode (at the Ic -VCE plateau) (Hu 2009, p. 293)

In saturated operation, the current I_C increases with the voltage V_{CE}, both emitter-base and base-collector diodes are connected in direct bias, and electrons are injected into the base by both diodes (Chapter 4, section 4.1.2 and Figure 4.3). The current of electrons injected by the collector-base diode increases with V_{CE} until saturation is reached (Figure 5.5). At "saturation", the "charge" (quantity of minority carriers, electrons for a base p) in the base is at its maximum: the base is said to be saturated.

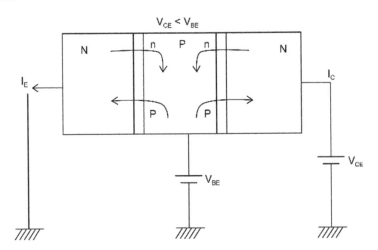

Figure 5.5. *Currents in a bipolar transistor*
operating in saturated mode, $V_{CE} < V_{BE}$

In active operation, the current I_C is independent of the voltage V_{CE}. The characteristic curve $I_C - V_{CE}$ shows a plateau (Figure 5.4). Each plateau corresponds to a voltage V_{BE}. In this regime, current I_C increases with voltage V_{BE} (Figure 5.4(b)). For a silicon-based transistor, below $V_{BE} < 0.65$ V, $I_C = 0$, the transistor does not conduct. The emitter-base diode $N^+ P$ is connected in direct bias V_{BE}. The P-N collector-base diode is reverse-biased, $V_{CE} > V_{BE}$. When a positive voltage V_{BE} is applied to the base of the emitter-base diode (forward bias), electrons from the N^+ emitter pass (are injected) into the base, becoming minority carriers. The majority of electrons injected by the emitter diffuse through the base and are absorbed by the collector, as a result of the high electric field at the base-collector junction (in reverse polarization) (electron flow). At the same time, a hole flow I_B flows in the opposite direction, from the base to the N region. The transistor is referred to as "bipolar", because both types of charge carrier flow.

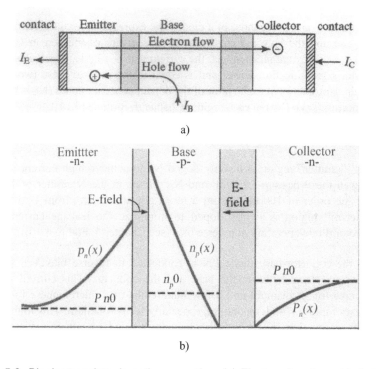

a)

b)

Figure 5.6. *Bipolar transistor in active operation. (a) Electron flow I_C and hole flow I_B (Hu 2009, p. 297); (b) minority carrier distribution: electrons n_p in base P, holes in emitter N (Lorenzini 2021). For a color version of this figure, see www.iste.co.uk/ vignes/silicon1.zip*

In active operation at the plateau of the characteristic curves, the current I_E, the sum of the currents I_C and I_B, depends only on the voltage applied to the base V_{BE} (Figures 5.4(b) and 5.6(a))[2]:

$$I_E = I_S \exp{(eV_{BE}/kT - 1)} \text{ with } I_E = I_B + I_C \qquad [5.1]$$

In the base and emitter regions, the minority carrier currents are diffusion currents (Chapter 4, section 4.3.2), electron-hole recombinations are very small given the very short distances ($L_B < 0.1 \ \mu m$) and the concentration profiles of the minority carriers, electrons in base B and holes in emitter E, are quasi-linear (Figure 5.6(b)).

Compare these linear profiles over very short distances with the corresponding profiles in a PN diode (Chapter 4, Figure 4.4(a)).

The currents I_B and I_C are those of a diode. The expressions for the electron current I_C in the base and the hole current I_B in the emitter are established in Chapter 4, formula [4.9]. As they are independent, the current gain $\beta = I_C / I_B$, which characterizes the transistor's amplification power and is equal to the ratio of these two currents. Current gain is proportional to the ratio of the dopant concentration n ($N_D = N^+$) to the dopant concentration p (N_A) in each region (Chapter 4, formula [4.14]):

$$\beta = I / I_{CB} = (\mu_n/\mu_P)^{1/2} \cdot \{N_D/N_A\} \qquad [5.2]$$

The N^+ "emitter" region is heavily doped N_D to achieve high current gain. For N_A doping in the P base of $10^{18}/cm^3$ and N_D doping in the N emitter of $10^{20}/cm^3$, gain is of the order of 100 (I_{Base} from 5 to 75 μA and $I_{Collector}$ from 1 to 10 mA). The "collector" region is lightly doped to minimize the leakage current of the base-collector diode operating in reverse bias (see Chapter 4, Figure 4.4(b)).

When the collector-base diode PN is connected in reverse bias $V_{CE} > V_{BE}$, a leakage current flows between the base and the collector. This current is a few microamperes for germanium and a few hundredths of a microampere for silicon, but increases rapidly with temperature, especially for germanium. This high leakage current for germanium is one of the factors that led to silicon's victory over germanium in bipolar transistors.

5.1.2.2. *Transit time of minority carriers*

In the case of the bipolar transistor N^+PN, polarized in the active operation (Figure 5.6(a)), the basic characteristic is the time required for electrons (minority

---

2 All the formulas presented in this chapter are established in the two basic works: (Sze 2002, Chapter 5; Hu 2009, Chapter 8).

carriers) to pass through the transistor base; in other words, the transit time of the minority carriers in the base.

This transit time is equal to the ratio of the quantity Q_B of minority carriers present in the base (area under the curve in the base (Figure 5.6(b)) to the current I_C (Hu 2009, p. 306):

$$Q_B / I_C = \tau_T \tag{5.3}$$

From the expression for the electron current I_C in the base and the charge Q, we establish that the transit time in active mode is proportional to the square of the length L_B of the base divided by the diffusion coefficient of the electron in the base (Hu 2009, p. 307):

$$\tau_T = L_B^2/2D_{n(p)} \tag{5.4}$$

For a base length of 70 nm and a diffusion coefficient of $D_{n(p)} = 10$ cm^2/s:

$$\tau_T = 2.5 \times 10^{-12} \text{ s} = 2.5 \text{ ps}$$

5.1.3. *Basic functions*

"Switching" and "amplification" are the two basic functions of bipolar transistors or MOSFETs (Chapter 6). Switching times and cut-off frequencies (frequency beyond which amplification no longer occurs) are the basic characteristics. Transistors have replaced triodes for these functions. These functions rely on the control of a current of electrons between emitter and collector for NPN transistors, which can be either amplified, interrupted or switched back on.

5.1.3.1. *Basic circuit operation*

These two functions – switching and amplification – are performed with the transistor biased in active mode. They are controlled by the voltage applied to the base V_{BE}. The current between emitter and collector I_C is an electron current through the base, which depends only on the voltage applied to the base V_{BE} (Figure 5.4(b)).

The basic circuit is the inverter (Figure 5.7(a)). It consists of a transistor and a load resistor R_D, controlling the voltage V_{CE}:

$$V_{CE} = V_{CC} - R_D.I_C \tag{5.5}$$

The operating point of the circuit F is determined by the intersection of the load line with the current plateau I_C or I_B corresponding to the applied voltage V_{BE} (active mode) (Figure 5.7(b)), which sets the current I_C and the voltage V_{CE}.

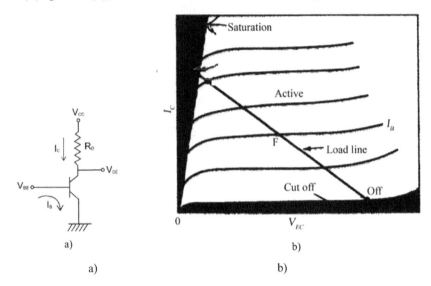

a)   b)

Figure 5.7. (a) Inverter circuit and (b) operating point, ON (F) in active mode (adapted from Sze (2002, p. 150))

Inverter circuit operation in active operation is controlled by the emitter-base diode, and therefore controlled by the voltage V_{BE}, it can be modeled as an "RC" circuit (see Chapter 4, section 4.1.3).

The expression of the capacity C is:

$$C = Q_B/V_{BE} \tag{5.6}$$

where Q_B is the "charge" of minority carriers present in the base, represented by the area under the curve in the base (Figure 5.6(b)).

The resistance R is defined by:

$$1/R = dI_C/dI_{BE} \tag{5.7}$$

where $1/R$ is the slope (tangent) of the characteristic curve $I - V_{CBE}$ at the operating point (Figure 5.4(b)).

Since capacity C is a constant, taking into account expressions [5.2] and [5.7], it is expressed by:

$$C = dQ_{BF}/dV_{BE} = \tau_T \, (dI_C/dV_{BE}) = \tau_T/R \qquad [5.8]$$

The RC circuit time constant τ_{RC} is equal to the transit time of the minority carriers through the base τ_T:

$$\tau_{RC} = \tau_T \qquad [5.9]$$

5.1.3.2. *Switching*

Switching is achieved by a voltage step V_{BE} applied to the base, which causes a current step I_B switching the transistor from the OFF state (blocked, $I_C = 0$) to the ON state, where the current I_C is defined by the point of intersection of the load line with the characteristic curve corresponding to the applied potential V_{BE} (I_B) (Figure 5.8(b)).

During the establishment of conduction, that is, for the duration of an OFF/ON switch, the electron current from the emitter splits in two. One fraction remains in the base, gradually building up the charge Q_B, and thus gradually saturating the capacitance C. A second fraction passes through the base. When the ON state is reached, the capacitor no longer plays any role, and all current flows through the resistor.

The switching time OFF/ON, that is, storage time of the Q_B charge of minority carriers, that is, of the circuit, is equal to three times the RC circuit time constant; therefore, it is three times the transit time of the minority carriers in the base:

$$\tau_C = 3 \, \tau_{RC} = 3 \, \tau_T \qquad [5.10]$$

where τ_{RC} is the RC circuit time constant.

For a voltage pulse (step type) applied to the base V_{BE} producing a current step I_B, the current I_C has the expression (Figure 5.8(a)):

$$I_C(t) = \beta \, I_{B0} \, \{1 - \exp(-t/\tau_T)\} \qquad [5.11]$$

It only reaches its value after a period of time $3 \, \tau_T$.

Destocking time depends on the speed with which electrons present in the base are eliminated by recombination with majority carriers, and therefore on the "lifetime" of these minority carriers in the base (see Chapter 4, section 4.1.3).

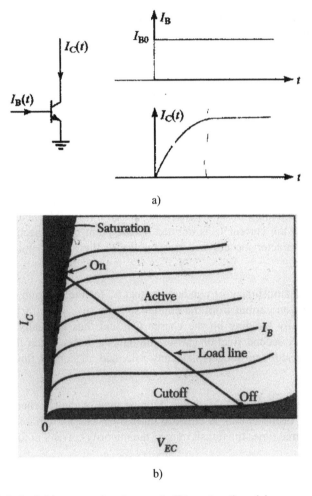

a)

b)

Figure 5.8. *Switching operation from cut off to saturation: (a) response; current I_C (t) to a voltage step V_{BE} producing a current step I_B; (b) graphical representation of ON/OFF switching operation (Sze 2002, p. 150)*

5.1.3.3. *Weak current amplification*

The transistor, as a component of a weak-signal amplifier circuit, can be used to increase the power of alternating signals by means of a DC power source.

The basic amplifier circuit is the inverter consisting of a transistor and a load resistor R_D controlling the amplified voltage V_{DS} (Figure 5.7(a)).

A weak signal is amplified by a transistor biased in active mode (Figure 5.9). The operating point is defined by the intersection of the load line with the step in the characteristic corresponding to the voltage applied to the base V_{BE} or I_B.

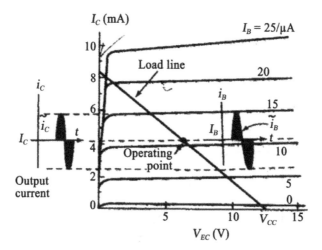

Figure 5.9. *Graphical representation of amplification. Sinusoidal signal to be amplified i_B; sinusoidal signal i_C amplified (Sze 2002, p. 147)*

The amplified current i_C is obtained by the graphical construction shown in Figure 5.9. The sinusoidal signal i_B (t) corresponding to the signal v_{BE} (t) to be amplified, whose extreme values are 2 and 15 µA (in the example given), produces an amplified sinusoidal signal i_C in the emitter-collector circuit between 2.0 and 6 mA.

In the common-emitter configuration shown (Figure 5.3(a)), the current gain i_C/i_B, which characterizes the transistor's amplification power, is equal to the ratio of the electron current I_C to the hole current I_B (formulas [5.2] and [5.3]); this is therefore proportional to the ratio of the dopant concentration n ($N_D = N^+$) to the dopant concentration p (N_A) in each region (formula [5.2b]):

$$\beta = I/I_{CB} = i/i_{CB} = (\mu_n/\mu)_P^{1/2} \cdot \{N/N_D\} \qquad [5.2b]$$

The "emitter" region is heavily doped with N_D^+ for high current gain. For an N_A doping in the P base of $10/cm^{183}$ and an N_D doping in the N emitter of $10/cm^{203}$, the gain is of the order of 100.

5.1.3.4. *Frequency response*

The gain of the transistor $\beta_F = I_C/I_B$ depends on the frequency of the signal applied to the base. At low frequencies, the current gain β_0 is independent of frequency. Above a certain frequency, the current gain decreases (Figure 5.10). A transistor's high-frequency operation is characterized by two "cut-off" frequencies.

The frequency f_β: frequency at which the gain $\beta < 0.707\ \beta_F$ (−3db).

The frequency f_T is defined as the frequency beyond which there is no further amplification, $\beta = 1$. It is expressed as a function of the RC circuit time constant circuit (formula [5.3]), that is, as a function of the transit time τ_T (formula [5.7]):

$$f_T = 1/2\pi\ \tau_T \text{ for } \beta = 1 \tag{5.12}$$

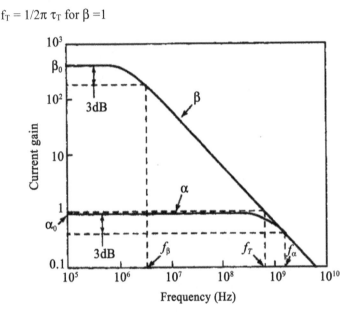

Figure 5.10. *Amplification: current gain as a function of operating frequency with frequency (Sze 2002, p. 148)*

5.2. Transistor technologies

"Advances in device technology were governed by properties of materials and the material scientists and metallurgists were the unsung heroes who made the industry viable" (Petriz 1962).

The manufacturing processes developed to achieve the required properties of purity and doping control for bipolar transistors have also been applied to the development of MOSFET transistors (see Chapter 6). Some of these most important developments are due to the persistence of metallurgists (chemists) present in the research laboratories of ATT (Bell Labs), General Electric, RCA, then Texas Instruments (TI), Fairchild Semiconductor.

With the bipolar transistor concept validated by Shive's experiments, the problem was to design a product that could be "industrially" manufactured.

For a bipolar transistor N^+PN, the idea is to create a stack of three layers, including a P intermediate layer of the smallest possible thickness, starting from an N substrate which will be the N collector, to create a P base of the smallest possible extension by p-type doping, controlling the doping and its extent in depth in the substrate, and then to create the N emitter$^+$ by additional n doping without destroying the monocrystallinity (Figures 5.18 and 5.19).

5.2.1. Single-crystal pulling: CZ process and mixed CZ-diffusion processes

5.2.1.1. Germanium bipolar transistor

The first germanium bipolar transistor was manufactured using the CZ process (grown junction process) by Bell Labs researchers in early 1950 (CHM 1951; Leamy and Wernick 1997).

Gordon Teal modified the single-crystal pulling plant (see Chapter 4, section 4.2.3.2) to allow controlled introduction of a dopant into the growing crystal (pill dropping). In the early months of 1950, Sparks, working closely with Teal and Buehler, began producing PN junctions using a germanium N crystal acting as the nucleus, immersed in a liquid bath of P (doped with Ga) (Teal 1955).

Recently converted to the merits of single crystals, Shockley became impatient and asked Sparks to make a PNP sandwich by adding the corresponding dopants while pulling a single crystal. In April 1950, Teal and Sparks used the CZ process to make a single-crystal germanium NPN sandwich. Starting with an n-doped growing crystal, by first introducing a limited quantity of gallium (p-type) into the liquid bath to form the base P, then antimony Sb (n-type) to form the emitter N, they obtained an NPN transistor with a base thickness of 30 mils (750 μm). The test of this sandwich as a transistor in April 1950 proved positive in that it demonstrated transistor operation; however, this result was of very limited interest as an amplifier of high-frequency signals (10–20 kHz), as the base was far too thick. The result aroused very little interest, and tests practically stopped for the rest of the year. Point-contact transistors performed much better up to 10 Mhz, which explains their use in calculators until 1954 (Shockley et al. 1951).

Studies and tests on the NPN transistor resumed in January 1951, under the impetus of Shockley following a request from the military for a special component (a detonator). By slowing down the pulling speed and stirring the liquid bath vigorously, Sparks succeeded in obtaining an NPN transistor with a P interlayer 1–2 mils (25–50 µm) thick. These components had a cut-off frequency of the order of 1 MHz, making them ideal amplifiers for AM signals (wavelength 300–600 m).

A mixed CZ-diffusion process with a grown diffused base was developed by TI to improve the frequency performance of germanium transistors by reducing the thickness of the base (Figure 5.11).

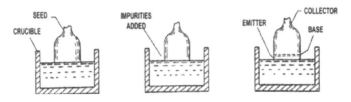

Figure 5.11. *Grown diffusion process for making a germanium bipolar transistor (Burgess 2011b)*

For a PNP transistor, the liquid germanium bath is p-doped, and the P collector is obtained by pulling (CZ process) (described in section 4.2.3). After growing a certain length of collector, n base and p emitter dopants are added to the bath at the same time. The P emitter is obtained by carrying on pulling. Simultaneously, the dopant n present in the emitter diffuses into the collector, creating a base of very narrow width. This method is used for germanium PNP and silicon NPN transistors, because in germanium, the n-type dopants (As, Sb) diffuse faster into the collector than the p-type dopant (In-Ga) (added for the emitter); whereas, in silicon, the opposite is true (see section 5.2.4).

5.2.1.2. Silicon bipolar transistor

During his last two years (1951–1952) at Bell Labs, Gordon Teal succeeded in manufacturing silicon single crystals using the CZ process, and PN junctions were realized by Pearson and Sawyer (1952) and Teal and Buehler (1952) (Chapter 4, section 4.2.4). "Immediately silicon became a rival to germanium" (Leamy and Wernick 1997).

In January 1954, Bell Labs chemist Morris Tanenbaum (Tanenbaum et al. 1955), using "high-purity" silicon supplied by DuPont, fabricated the first NPN silicon bipolar transistor using the process developed by Sparks and Teal for germanium transistors. Its operating performance was superior to that of the germanium transistor (CHM 1954a). However, the quality of the silicon was such that the lifetime of minority carriers through the base was too short, even for 1 mil bases (25 μm).

5.2.2. Philco process

The Philco process, developed by the company of the same name in 1953, is made up of an electrolytic etching of a germanium N wafer placed between two aligned jets of an indium salt-based electrolyte, until a fixed thickness is reached. Then, by reversing the polarity of the supply, indium (a p-type dopant) is deposited on both sides of the thinned wafer, forming two In-Ge metal junctions. By heat treatment, P regions are produced by alloying Ge-In. This produces a PNP transistor (Bradley 1953).

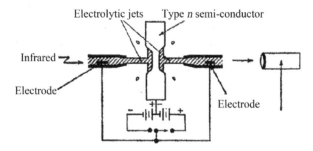

Figure 5.12. *Philco process: etching and electroplating (Dezoteux and Petit-Jean 1980, p. 40)*

This so-called "surface barrier transistor" (SBT) was the basic component of the first computers: Philco Transac 2600 in 1956, LARC in 1960. The first car-mounted transistor radio was developed and produced by Chrysler and Philco in 1955.

5.2.3. "Alloying" process

This complex process, developed solely for germanium transistors due to germanium's low melting point, was developed for transistors used in computers (including IBM). It is presented in some detail to show the difficulty of manufacturing such transistors.

In 1949, Hall of General Electric[3] produced a germanium PN diode with high-power rectifier characteristics by bringing together a germanium wafer (very

---

3 For a presentation of the work carried out by General Electric, see Burgess (2011a).

thin) with an indium pellet (p-type dopant) on one side and an antimony pellet (n-type dopant) on the other. Since both elements have low melting points, during heat treatment of the assembly, the germanium wafer disappeared, forming an alloy with the two wafers, two P and N regions, giving the diode rectifying properties (Hall and Dunlap 1950; Hall 1955).

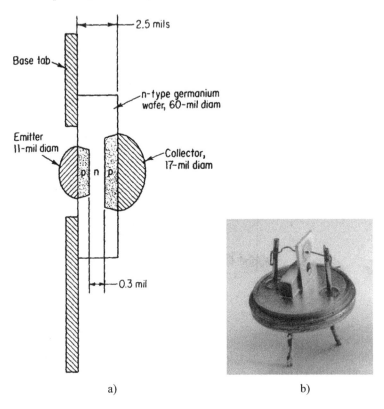

a)   b)

Figure 5.13. *(a) Schematic representation of the "alloying" process (Pankove 1961); (b) IBM 083 NPN germanium transistor produced by alloying (Gardner and Dill 2010). For a color version of this figure, see www.iste.co.uk/vignes/silicon1.zip*

In 1951, John Saby of General Electric manufactured a germanium PNP bipolar transistor using this process (Saby 1961). At the same time, Jacques Pankove of RCA, inspired by Hall's work, produced a bipolar transistor using this process. The manufacturing process for this type of transistor, described in Pankove's (1961) patent, is shown in Figure 5.13(a): on a thin, single-crystal germanium N(Sb) wafer with a thickness of 75–150 µm, a spherical indium (p-type dopant) dot with a diameter of 25–50 μm is successively positioned on each side. The collector dot is applied first. It is stuck to the wafer by heating (firing) at 250°C for 1 min, causing it to melt and spread by wetting (T_f indium: 156°C), the extent of which is controlled; control of this spreading is an important factor in the uniformity of the results. The emitter dot is then applied to the other side using the same process. The whole assembly is heated to 400–500°C for 10–20 min, treatment during which both indium pellets melt. The liquid indium reacts with the solid germanium to form an alloy. Two zones of indium-germanium alloy are formed, producing a thin layer of p-doped emitter and a layer of p-doped collector, on either side of the N germanium wafer.

Together, they form the PNP bipolar transistor. The solubility of indium in germanium at 500°C is of the order of 3×10^{18} atoms/cm^3, that is, very low. The germanium in the base, being soluble in liquid indium, dissolves in both dots (the maximum solubility of germanium in indium is 10%). The faces of the germanium crystal are oriented <111>, the direction of slow dissolution. An In-Ge alloy is formed on either side of the germanium base. This operation controls the remaining thickness of the N-type germanium layer that forms the base of the transistor, hence the name of the process. During cooling, the two molten regions recrystallize, reforming a single crystal with the germanium base by directional solidification: a prerequisite for high-performance NPN transistors. To operate at very high frequencies, the thickness of the inner layer (the base) must be reduced to 10–20 μm. In fact, this type of transistor could not be used to produce transistors with very high cut-off frequencies because it was impossible to control the thickness of the base.

TI (Figure 5.23), RCA, Raytheon, GE and Motorola began producing germanium transistors using this process. This type of transistor was then developed by IBM in 1952 (Figure 5.13(b)) to equip its future computers: the IBM 608 and 1401 models. Telefunken also developed this type of transistor.

5.2.3.1. Drift transistor

A variant of the germanium transistor obtained by the alloy process was developed by RCA in 1956 (2N747)[4], initially designed in Germany by Herbert Krömer (Krömer was to win the Nobel Prize for other work) (Knight 1957). This drift transistor is manufactured by alloy-diffusion, but the diffusion doping of the N-base is not uniform: strong on the emitter side and weak on the collector side to create an internal (electric) field that accelerates minority charge carriers as they pass through the base. Cut-off frequencies in excess of 300 MHz were achieved (Schwartz and Slade 1957).

---

4 For a detailed presentation of RCA's work, see Burgess (2008b).

5.2.4. *CVD diffusion process*

5.2.4.1. *The process*

This CVD (chemical vapor deposition) diffusion process, developed for the production of silicon solar cells (Volume 2, Chapter 4, section 4.5.3) and NPN bipolar transistors, first in germanium, then in silicon, was also the basic technology for MOSFET transistors (in silicon).

This process for producing thin n-doped surface zones or layers on a P substrate (or vice versa) by diffusion consists of subjecting the surface of a wafer to a heat treatment in the presence of a gaseous phase comprising of a compound of an n (phosphorus) or p (boron) dopant, which, by decomposing or reacting with another reagent in contact with the surface, thus produces a thin deposit of dopant. This dopant, having a certain solubility in the semiconductor, diffuses inside the wafer. This is a mixed process, consisting of a CVD step for depositing a dopant on the surface of the wafer (Chapter 4, section 4.2.2.1), and solid-phase diffusion of the dopant into the wafer (simultaneous or subsequent).

Since the solubilities and diffusivities of these dopants in silicon and germanium are low, by controlling the temperature, composition and vapor pressure of the gas phase, we can control the deposition reaction, and therefore the thickness of the dopant layer and consequently the penetration depth and concentration profile of the dopant (Figure 5.14); and hence the thickness of the N layer on the P wafer. The result is a P-N junction, or the base of a solar cell (Volume 2, Chapter 4).

For silicon, boron (p) is the basic element for forming P layers, while arsenic and phosphorus (n) are used to form N layers. These dopants can be introduced from a variety of sources: solid sources (BN, As_2O_3, P_2O_5), liquid sources (BBr$_3$, AsCl$_3$, POCl$_3$) and gaseous sources (B$_2$H$_6$, AsH$_3$, PH$_3$). For example, P doping of silicon is achieved by the following reactions:

$4POCl_3(l \rightarrow g) + 3O_2(g) \rightarrow 2P_2O_5 (s)$ (deposited on substrate) $+ 6\ Cl_2(g)$

$2P_2O_5 (s) + 5\ Si(s) \rightarrow 4P (s) + 5\ SiO_2$

$P(s) \rightarrow P$ (diffusion in Si) (N layer)

Bell Labs researchers' work on this process began with Fuller's work in 1951 (on a problem encountered with germanium, namely that during heat treatment at 500°C, germanium N was converted into germanium P). This conversion was due to contamination by Cu. The diffusivity of copper in germanium or silicon was very high. This result led to work on the diffusion of other p and n contaminants and dopants in germanium and silicon. Fuller showed that in germanium, the n-type dopants As and Sb diffuse much faster than the p-type dopants Ga and In, while in silicon, the opposite is true. The diffusion speed of p-type dopants (B, Al) is 10–100 times greater than that of n-type dopants (Sb, As), and the surface concentration of n-type dopants is 10–100 times greater than that of p-type dopants. So, by simultaneous or sequential diffusion of n (P, As) or p (B, Al) dopants in N silicon, we obtain an inner p layer as a result of faster boron diffusion, and an outer N layer rich in P or As, constituting the emitter as a result of donor accumulation (Figure 5.14; Fuller 1962). In December 1954, Lincoln Derick and Carl Frosch perfected the double diffusion process (Derick and Frosch 1957b).

The double-diffusion doping process, which enables the two successive layers (base and emitter) to be produced on an N collector wafer, proved to be the best way to obtain NPN bipolar transistors with base thicknesses of less than a fraction of 1 μm, and thus to achieve high-frequency signal amplification performance of up to gigahertz.

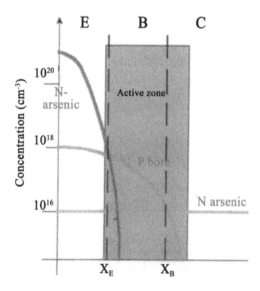

Figure 5.14. *Double diffusion process: emitter (E) and base (B) in a substrate (C). Concentration profiles are shown. For a color version of this figure, see www.iste.co.uk/vignes/silicon1.zip*

The development of this technology took a long time because of the impurities in the gas phase, which, when introduced simultaneously with the dopant even at extremely low levels, had a great effect on electrical properties, and whose levels must be lower than those of the dopant, which are themselves very low, from 10^{15} to 10^{17} atoms/cm^3.

The second problem was how to make ohmic contacts with these different layers.

5.2.4.2. *History*

The diffusion doping process was proposed by Jack Scaff (the latter being the director of the Materials Research Laboratory at Bell Labs) and Henry Theuerer at the end of 1947 to produce P-N junctions with a photoelectric effect.

Bell Labs were undoubtedly the pioneers in the development of the CVD-diffusion process (from a gas phase) applied to the manufacture of silicon solar cells in 1954 and to the production of bipolar transistors (NPN) in germanium and silicon from 1954[5]. It should be emphasized that this development is due to the persistence of Calvin Fuller.

In 1952, Fuller began diffusion doping tests on germanium, then on silicon. At the end of 1953, Fuller used boron diffusion (p-dopant) on silicon N wafers to produce large-area P-N junctions with a strong photoelectric effect: conversion efficiencies of up to 6%. In March 1954, Fuller filed his first patent on the diffusion process for doping silicon with boron (Fuller 1962). On April 26, 1954, Bell Labs announced the manufacture of silicon solar cells using the diffusion doping process (Charpin et al. 1954).

The first transistors manufactured by Bell Labs used a mixed process: "CVD-diffusion" for the base and "alloy" for the emitter:

1) In late 1954, Lee of Bell Labs (Lee 1956), produced the first germanium PNP transistor using the CVD-diffusion process to produce the base, and the alloy process to produce the emitter of a PNP transistor (Figure 5.15). On a germanium P wafer, an n-type layer approximately 1 μm thick, forming the transistor base, is produced by arsenic "diffusion". An aluminum metal deposit of limited extent is then deposited of the base surface, which by "alloying", forms the emitter. The ohmic contact of the base (in gold) is also made by metal deposition on a very limited area of the base. To produce the lowest possible base-collector junction, the periphery of the wafer is then etched. Such PNP bipolar transistors with a "diffused" base could amplify signals with a frequency of 500 Mhz, almost 10 times higher than the best transistors available at the time (CHM 1954b).

---

5 For a comprehensive overview of research, development and production by Bell Labs and its parent company Western Electric, see Burgess (2010).

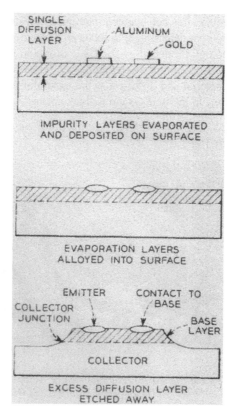

Figure 5.15. *Schematic representation of the steps involved in making a bipolar transistor (Sparks and Pietenpol 1956) (see Figure 5.18)*

2) In late 1954/early 1955, Nick Holonyak prototyped a silicon PNP bipolar transistor (Figure 5.16). On a P silicon wafer, an N layer is produced by phosphorus diffusion, forming the base of the transistor. The P layer, constituting the emitter, is then deposited with aluminum and diffused by heat treatment. These transistors performed poorly due to the low solubility of aluminum in silicon, and the base doping was too high to achieve an efficient emitter-base junction (Holoniak 2007).

3) In March 1955, Tanenbaum and Thomas, using simultaneous "double diffusion" of aluminum and antimony in a silicon N substrate, produced prototype transistors with a base thickness of 2 μm. This prototype was the first good silicon transistor made by the "double diffusion" process (Tanenbaum and Thomas 1956).

According to Holonyak, this achievement convinced Jack Morton, head of Bell Labs' transistor development department, to focus all of the department's activity on silicon transistor technology produced by the CVD-diffusion process. According to Riordan, Morton boldly decreed that, from that day on (end of March 1955): "It will be with silicon as the material and diffusion as the technology that the development of the transistor and diode in the Bell system will take place".

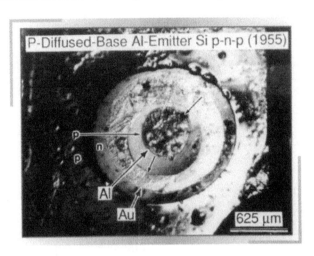

Figure 5.16. *"Mesa" transistor (produced in early 1955) on a silicon disk P using the diffusion-alloy process (see Figure 5.18) (Holonyak 2007)*

In January 1956, Bell Labs revealed to its licensees the technologies it had developed to produce its transistors.

According to Morton in 1958 (Morton and Pietenpol 1958), the following results were obtained:

– diffused-base germanium transistors with cut-off frequencies up to 1 GHz and switching times of 10^{-8} s (10 ns);

– diffused-base silicon transistors with cut-off frequencies from 50 to 100 Mhz and switching times of 10^{-7} s.

5.2.5. *"Ion implantation" process*

As with the diffusion process, it is to Bell Labs that we owe the recommendation and early development of this technology, and, in particular, to two metallurgists, alongside Gordon Teal namely Russell Ohl to whom we owe the discovery of the rectifier effect and the photoelectric effect of PN diodes (see Chapters 2 and 3) (Ohl 1952, 1956; Shockley 1957).

The ion implantation process consists of implanting dopant atoms into the top layer of a substrate by bombarding the surface with ions of the element accelerated by an electric field. This process enables quantitative control of penetration depth (by ion energy) and of the element's vertical composition profile by ion flow rate. It is a highly directional process, enabling selective implantation in well-defined areas, using masks with very sharp borders (Figure 5.17). Penetration depth is of the order of 100 nm. However, ion bombardment introduces defects into the substrate through the creation of gaps and the formation of dislocations by clusters of gaps. Hence, the need for annealing to restore lattice perfection.

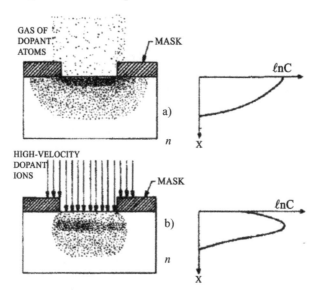

Figure 5.17. *Doping processes: (a) by diffusion; (b) by ion implantation of delimited zones; concentration profiles (Jackson 2005, p. 381)*

5.3. "Mesa" and "planar" bipolar silicon transistors

5.3.1. *Oxide masking*

The oxidation masking process developed in 1955, coupled with the use of photolithography, enabled the production in 1958 of mesa and planar bipolar silicon transistors of well-defined dimensions, using the diffusion doping process.

In early 1955, while developing the CVD-diffusion process (section 5.2.4), Bell Labs researchers encountered the problem of surface degradation (with pitting, erosion and even destruction) of the silicon substrate, following an increase in temperature (requested by the researchers) for the silicon phosphorus diffusion treatment.

In a test conducted by Lincoln Derick and Carl Frosch, the hydrogen gas transporting the volatile dopant compound into the diffusion reactor "caught fire" and water vapor entered the reactor, producing a thin layer of silica on the silicon substrate, which proved to be a highly stable, pinhole-free protective layer (very few metals form pinhole-free oxide layers) (Derick and Frosch 1957b).

Derick and Frosch then showed that the oxide layer could be used as a selective mask to define doping zones (emitter and receiver zones of a bipolar transistor or source and drain zones of a MOSFET transistor) on the substrate (see Figure 5.20 and Volume 2, Chapter 1, and Figure 1.20).

They showed that certain dopants could not diffuse through the oxide layer (the latter constituting a barrier to their penetration into the substrate), and thus that by etching windows in the oxide layer, certain well-defined zones could be doped. Dopants (P, As, Sb, n-type) do not diffuse through the oxide layer in the presence of an oxidizing atmosphere. Boron, on the other hand, diffuses through the oxide layer in the presence of hydrogen or water vapor.

5.3.2. *Mesa structure*

In 1958, Fairchild Semiconductor[6] was the first company after Bell Labs to use the oxide masking process to manufacture a mesa-structured NPN transistor in silicon, where the base and emitter stacked were carried out by diffusion. The extension of the emitter was controlled by a window etched into the oxide layer after the base had been fabricated (Figure 5.18). The 2N696 NPN mesa transistor on a Si-N substrate (doped with Sb) was designed by Gordon Moore (identical to the Bell transistor 2N560) and the first PNP mesa transistor 2N1131 on a Si-P substrate (B-doped) was produced by Jean Hoerni.

The first NPN bipolar transistor in silicon mesa (2N706), doped with gold (electron-hole recombination center manufactured by Fairchild in 1960), had a switching time of 16 ns.

---

6 In this article, Gordon Moore describes in detail all the technical and technological problems encountered by Fairchild in the industrial production of a silicon transistor using the double diffusion process (Moore 1998).

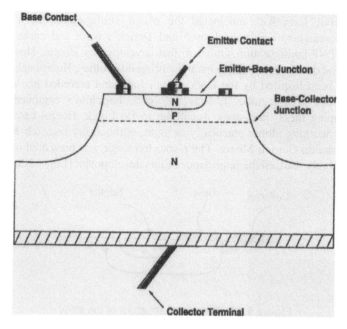

Figure 5.18. *Early bipolar mesa transistor*
(see Figure 5.15) (Moore 1998)

5.3.3. *Planar structure*

It was the invention of the "planar" structure in 1959 that made it possible to produce silicon-based bipolar transistor integrated circuits, according to Holonyak (2007).

According to Lojek, "The planar process was, after the invention of the junction bipolar transistor, the most important invention of microelectronics" (Lojek 2007, p. 120).

The silicon NPN bipolar transistor of mesa structure had one drawback: the connection wires prevented the creation of complex circuits.

In 1959, the invention of the planar structure solved the problem. PNP transistors are made directly on the silicon surface, with emitter and base successively diffused into windows in the oxide layer, and connections made on one side of the transistor by depositing metal wires on the surface (Figure 5.19).

History credits Jean Hoerni of Fairchild Semiconductor for the invention of the planar structures of bipolar NPN transistors. In fact, as early as 1955, Frosch and

Derick at Bell Labs had anticipated the planar configuration of a transistor. Bell Labs researchers continued Frosch and Derick's work and came up with a planar-type PNP configuration similar to that developed by Hoerni. However, they abandoned the development due to manufacturing difficulties. Hoerni acknowledged that he had been inspired by the work at Bell Labs and recorded his ideas in his laboratory book on December 1, 1957. However, Fairchild's engineers were too busy developing mesa transistors. According to Bo Lojek, Hoerni carried out his experiments working alone, practically at night, without any research budget, and without informing Gordon Moore. The planar transistor was presented on March 4, 1959, but nobody realized the importance of this development (Lojek 2007, p. 123).

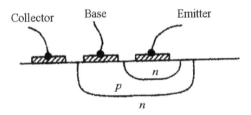

Figure 5.19. *The planar structure of the NPN bipolar transistor (Lojek 2007, p. 123)*

The sequence of operations for building an NPN bipolar transistor with a planar structure developed by Hoerni is shown in Figure 5.20. The structure is made of a single-crystal substrate with low n-doping (32). It is covered by a layer of oxide fabricated in situ (31). The base (36) (an island in the collector) is produced by diffusion of a p-dopant through an open window in the oxide layer. A new oxide layer and a new window in this layer are created. The emitter (39) (island in the base) is created by new n+ (strong) doping diffused into this window. The three metal contacts (44, 46, 47) (aluminum) to the collector, base and emitter are made at corresponding openings.

The successive windows were photolithographed. Photolithography is presented in Volume 2, Chapter 1.

The critical dimension for transistor switching speed of the transistor is the distance between emitter and collector, that is, the path traveled by the electrons injected by the emitter and captured by the receiver. For bipolar transistors with a planar structure, it is vertical (Figure 5.19) and controlled by the process (time and temperature) used, that is, successive diffusion of dopants to create the base, and then the emitter.

The problems encountered by Hoerni in this development were due to the presence of pinholes in the masking oxide layer, which forced Hoerni to develop a sophisticated oxidation process (presented previously).

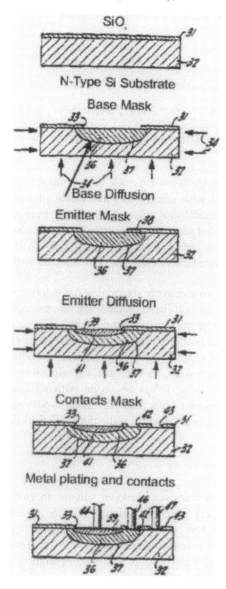

Figure 5.20. *Main processing steps of the planar NPN transistor (Hoerni 1962a)*

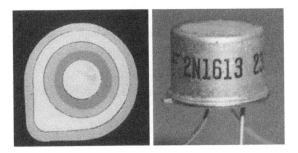

Figure 5.21. *The first 2N1613 silicon NPN planar transistor by Jean Hoerni (Ward 2004). For a color version of this figure, see www.iste.co.uk/vignes/silicon1.zip*

It was the masking oxide layer that made this configuration possible. As this layer could only be produced with silicon, this innovation (the planar configuration for transistor manufacture) established silicon's superiority over germanium as the basic material for transistors, and later integrated circuits. Previously, silicon had been used for expensive components and germanium for cheap ones. With the planar structure of transistors, silicon integrated circuits became commercially viable.

Silicon's success as an electronic material is not solely due to its semiconductor properties. The elements that have determined its triumph are its temperature resistance (compared to germanium) and the quality of its natural oxide: silica. This stable oxide is one of the best electrical insulators known (breakdown field in excess of 10 million volts per cm). It chemically protects the silicon surface from any physical or chemical aggression. Its interface with silicon is abrupt and defect free. No other known semiconductor has an oxide with all these qualities. It is the properties of the silicon oxide layer that have made it possible to produce NPN and MOSFET transistors in planar structure using the CVD-diffusion process.

5.3.4. *Silicon epitaxial layer*

For an N^+PN bipolar transistor made of silicon, an epitaxial layer N-doped is produced using the CVD process, which forms the N collector, on a heavily N^+-doped substrate of low resistivity (which enables ohmic contact with the aluminum connecting wires) (see section 2.1.2). In the epitaxial layer, the P and N^+ zones$^+$ are created successively, as in the configuration shown in Figure 5.19. This structure greatly increases switching speed.

Control Data Corporation (Cray) awarded Fairchild a development contract to produce a silicon bipolar transistor with even higher performance (switching time

< 3 ns) than the 2N706 transistor with a switching time 16 ns (section 5.3.2). Jean Hoerni developed such a transistor (2N709) in 1961, combining the following developments: planar structure, epitaxial layer and gold doping (Hoerni 1962b). The central processing unit (CPU) of the CDC 6600 computer (Cray) released in 1964 was made up of 600,000 NPN transistors (2N709).

For a MOSFET transistor, an undoped epitaxial layer on the substrate is used for local ion implantation using masks to create the "inversion channel" p-doped and to control the threshold voltage (Chapter 6, section 6.2.2.1).

For both types of transistors, deposition of an epitaxial layer of lightly doped silicon on the single-crystal substrate became an essential step in the production process.

Once again, it was Bell Labs researchers who recommended and developed the epitaxial layer deposition process. In 1951, Gordon Teal (Teal again), assisted by Howard Christensen, developed the process for depositing epitaxial layers of germanium on a germanium substrate (Christensen and Teal 1954). In 1960, at the request of Ian Ross (already mentioned), Henry Theuerer perfected the CVD process for the depositing process to deposit an epitaxial layer of silicon on a single-crystal silicon substrate (Theuerer et al. 1960).

However, the industrial realization of this type of deposition posed a number of problems. Deposition is achieved by decomposing a silane or silicon chloride compound on the surface of the single-crystal substrate at high temperature, so that the mobility of the silicon atoms be sufficient for them to move to occupy sites, extending the underlying single-crystal structure of the silicon atoms. A perfectly clean interface is therefore necessary to achieve epitaxial deposition. The substrate surface must therefore be cleaned prior to deposition. For this etching operation, hydrochloric acid HCl is used, which is a source of particularly harmful impurities, such as transition metals resulting from acid etching of the reactor channels (whose role in shortening the lifetime of minority carriers we have already seen).

Fairchild (2N914) and TI were quick to adopt this technology for integrated circuits, based on planar NPN bipolar transistors in silicon.

5.4. Industrial developments

The industrial development of bipolar transistors kept pace with technological developments in manufacturing, which enabled ever shorter switching times, which in turn led to the development of commercial computers.

The first germanium bipolar transistor was manufactured by Bell Labs in April 1950 using the CZ grown junction process.

The invention of the bipolar transistor was announced in the press on July 4, 1951. As soon as Shockley's patent on this transistor was granted on September 25, 1951 (filed June 26, 1948; Shockley 1951a), Western Electric began licensing the right to manufacture components and transistors for $25,000. For its licensees, Bell Labs organized a Transistor Technology Symposium in September 1951.

Sensing the importance of these inventions, industrial developments were then carried out by vacuum tube companies that had acquired the Bell license, such as General Electric, Sylvania, RCA, Philco, CBS, Raytheon, Motorola, General Transistor/General Instruments (supplier of transistors for CDC 1604 (Cray) computers), Telefunken, Siemens, Sony; some of these were already heavily involved in the development of germanium diodes, such as Sylvania, General Electric and computer manufacturing companies like IBM[7].

Two companies, TI (Texas Instruments) and Fairchild Semiconductor, emerged and took over from Bell Labs in both the development and industrial production of silicon transistors, going so far as to be the inventors of "integrated circuits" (Volume 2, Chapter 1, section 1.2).

TI, created on January 1, 1951, grew out of an oil exploration company (GSI) which, under the impetus of Pat Haggerty, converted to a component manufacturer, after buying the manufacturing license and hiring Gordon Teal from Bell Labs in late 1952 (the father of the development of the CZ process for pulling single crystals of germanium and silicon). Teal succeeded in manufacturing the first NPN bipolar transistor in silicon-silicon using the CZ process in 1954.

Fairchild Semiconductor Corporation was created in 1957 as a self-sufficient division of Fairchild Camera and Instrument, which provided the initial funding for its development, by defectors from the company William Shockley had founded after his departure from Bell Labs. Fairchild produced the first bipolar silicon transistor with a mesa in 1958 and was the inventor of the first planar-structure transistor.

All of these companies had to set up all the transistor manufacturing operations from single crystals.

---

7 The development of these companies (e.g. General Electric sold its IT division to Honeywell in 1970; Honeywell sold its IT division to Bull in 1986) is described in *Computer History Museum: Companies* (see: www.computerhistory.org).

From the 1950s onwards, innovation in "computing" was transferred to the business world under the impetus of these companies.

These companies benefited from military and civilian programs to build ballistic missiles and earth satellites (Explorer 1).

5.4.1. *Germanium bipolar transistors*

The first bipolar transistor developments were focused on germanium transistors due to the relative ease of manufacturing.

In 1953, over 1 million germanium transistors were manufactured; in 1955, over 3,500,000; and in 1957, 29 million. From 1960 to 1965, sales of bipolar germanium transistors rose from 119 million to 334 million units; during the same period, sales of silicon transistors rose from 9 to 273 million units. Production of germanium bipolar transistors continued until 1979: 2 billion transistors were sold.

In 1951, the Air Force entrusted Bell Labs with the development of the TRADIC (Transistorized Airborne Digital Computer), which was entrusted to Felker. This resulted in the production of four successive TRADIC computers, of which the Leprechaum version, operational in 1956, was the first fully transistorized computer based on logic circuits made of bipolar germanium transistors (Felker 1951).

The first consumer product to feature a transistor (and two vacuum tubes) was manufactured by Sonotone in 1952. Raytheon supplied three CK718 transistors for a fully transistorized hearing aid manufactured from 1953 by the Maico company (CHM 1952).

The first transistors produced using the CZ process, by Western Electric and TI, were marketed in March 1953 (Figure 5.22). The first application was for hearing aids; TI received an order for 7,500 transistors from the Sonotone company in October 1953, which implied that TI had developed the CZ single-crystal pulling process. The only truly commercial application was the production of transistors for the first portable radios, the Regency TR-1, in 1954 by TI. 100,000 of these were sold within a year (CHM 1952).

This commercial success enabled TI to become an IBM supplier from December 1957. TI continued to manufacture this type of germanium transistor worldwide until 1964.

**Western Electric
Types 1858 and 1859**

TYPE
Germanium NPN Grown Junction Transistor

USAGE
Experimental Types for Voice
and Carrier Frequency
Telephone Equipment

DATE INTRODUCED
1953

Figure 5.22. *Germanium bipolar transistors manufactured by the CZ process in 1954 (Ward 2016). For a color version of this figure, see www.iste.co.uk/vignes/silicon1.zip*

In 1953–1954, Philco Corporation designed and launched the manufacture of a transistor: the SBT, operating at very high frequencies (20–30 MHz) (section 5.2.2). In 1957, 10 years after the discovery of the transistor, the first fully transistorized computer with SBT transistors in germanium was built: the TX-O, built by MIT Lincoln Labs. The first commercial computers were built in 1957–1958 by the Philco company: the Transac 2000, 2600, using its transistors. In 1960, the first "LARC supercomputer" was built by UNIVAC (IBM's competitor at the time) for the Lawrence Radiation Laboratory, using these SBT transistors.

**TEXAS INSTRUMENTS
R212**

TYPE
Germanium PNP Alloy Junction Transistor

USAGE
Polaris Missile Guidance Computer

DATE INTRODUCED
Late 1950s

Transistor Size (1/4" Diameter X 1/4"H)
Standard TO-5 Case
Date Codes 7111 (1971) and 6741A (1967)

Figure 5.23. *Texas Instruments R212 transistors manufactured by "alloying" (Ward 2006). For a color version of this figure, see www.iste.co.uk/vignes/silicon1.zip*

The next technological development was the "alloy" process for germanium transistors developed by General Electric in 1951, and simultaneously and independently by RCA. TI (Figure 5.23), RCA, Raytheon, GE and Motorola all began producing transistors using this process. This type of transistor was then developed by IBM in 1952 (Figure 5.13). Telefunken also developed this type of transistor.

One of the first military uses of this transistor was on the Polaris ballistic missile, for which the program was launched in early 1956 and commissioned in July 1960; this equipped the inertial guidance computer.

The first germanium NPN transistors for future IBM 608 computer models, marketed from 1957 (32 units sold), were manufactured using the "alloy" process (Figure 5.23). These computers operated with DTL (diode-transistor logic) circuits of discrete components (Volume 2, Chapter 1). IBM 1401 computers were equipped with transistors designed and manufactured by IBM in 1954. The IBM 1401 computer's arithmetic-logic unit featured 10,600 bipolar germanium transistors and 13,200 germanium point-contact diodes. Note that 15,000 computers were manufactured in the 1960s. IBM built an automatic production line capable of producing 3,600 transistors per hour. However, as IBM was not a manufacturer of electronic components, it subcontracted the production of these transistors to TI, transferring the production line to TI (Gardner 2010). TI continued to produce transistors using this technique until 1979, when 2 billion transistors of this type were sold.

These transistors, also manufactured by General Transistor Corporation at Seymour Cray's request, were used in the logic circuits of Cray's CDC 1604 computers in 1960: the first large-scale computer for scientific computing.

However, the type of transistor produced by the alloy process did not allow manufacturing transistors for very high frequency reception, as it was impossible to control the thickness of the base. A new transistor model was developed by RCA in 1956 (2N747): the drift transistor designed by Krömer (presented in section 5.2.3). The 7000 series computers, IBM 7030 Stretch and IBM 7090, were equipped with germanium drift transistors manufactured by IBM (IBM type 065), some of which were produced by TI.

The diffusion doping process, which was to become the most important manufacturing process (it was superseded by the ion implantation doping process), developed by Bell Labs for the production of silicon solar cells, was applied to the production of germanium and silicon transistors in 1954. In January 1956, Bell Labs revealed the technologies developed for these transistors to its licensees.

The first germanium transistor of this type manufactured by TI (2N623) in 1956 and their successors (2N1142) had a cut-off frequency of 750 Mhz. From 1960, TI was IBM's supplier of such transistors (germanium diffused base switching types).

Motorola and RCA manufactured millions of "diffused base" germanium transistors until the mid-1960s.

5.4.2. *Silicon bipolar transistors*

In May 1954, TI announced the production and sale of silicon NPN transistors manufactured using the CZ process (type 900), with a base length of 0.5 mil (12.5 μm) (Figure 5.24). Indeed, in December 1952, Gordon Teal had left Bell Labs for TI, with the mission of developing a silicon transistor for military applications. The first 150 silicon transistors produced by this process secured TI's future[8]. TI had no competitors until 1958. The first commercial transistors were produced in 1954: the TI 9000 series.

In 1954, several companies simultaneously developed the manufacture of silicon transistors: RCA (SX-152) and Hughes Aircraft (alloy junction transistor), Raytheon (CZgrown junction transistor) and Philco (silicon SBT).

Following the development of the diffusion doping process by Bell Labs in 1955, TI was the first company to market a mesa-structured silicon PNP bipolar transistor structure (2N389) in 1957, followed by Western Electric (2N560).

Transistor Size (1/4"H X 1/4"OD)
TI 2N336 Date Code 815 (1958 Week 15)
GE 2N333 Date Code 752 (1957 Week 52)

TI and GE 2N33X

TYPE
Silicon NPN Grown Junction

USAGE
General Purpose Industrial and Military

DATE INTRODUCED
1957

Figure 5.24. *Silicon bipolar transistors manufactured using the CZ process (Ward 2005). For a color version of this figure, see www.iste.co.uk/vignes/silicon1.zip*

In 1958, after Bell Labs, Fairchild was the first company to use the oxide masking process to manufacture an NPN transistor in silicon mesa structure by

---

8 Teal (1976) quotes *Fortune* magazine, November 1961: "The silicon transistor was a turning point in TI's history, for with this advance it gained big head start over the competition in a critical electronic product; there was no effective competition in silicon transistors until 1958. TI's sales rose almost vertically; the company was suddenly in the big leagues".

double diffusion (2N696 and 2N697) (Moore 1998). The dimensions of the Fairchild mesa transistor (2N696) were identical to those of the Bell Labs transistor (2N560).

Thanks to its temperature stability, this transistor was used in Minuteman missiles. In August 1958, Fairchild produced such transistors for IBM, for use in an on-board computer.

Thus, in view of the performance of the first NPN silicon bipolar transistor with high switching speed (16 ns), manufactured by Fairchild in 1960 (2N706), doped with gold (which was a version of the first bipolar transistor manufactured industrially using the diffusion process), Control Data Corporation (Cray) awarded Fairchild a development contract for an even higher-performance bipolar transistor (<3 ns), to be used in Cray's first CDC 6600 computer. Released in 1964, this computer used 600,000 gold-doped silicon NPN transistors (2N709) with a mesa structure in the CPU (Fairchild documentation).

The first planar-structure silicon transistor was manufactured by Fairchild (2N613) in 1959. From 1961 onwards, transistors manufactured by TI and Fairchild were of planar structure.

The inventions of the silicon oxidation masking process and the planar structure of transistors established the superiority of silicon over germanium as the basic material for transistors, then integrated circuits, and finally microprocessors. Previously, silicon had been used for expensive components and germanium for cheap ones. With planar transistors, silicon integrated circuits became commercially viable.

5.4.3. *Vanguard 1, Explorer 1 and 3 satellites*

The launch of the Russian Sputnik satellite in October 1957 was followed by the launch of the Explorer 1 satellite on February 1, 1958, designed and built by Caltech's Jet Propulsion Laboratory, and then by the launch of the Vanguard satellite on March 17, 1958, the first solar-cell-equipped satellite (see Volume 2, Chapter 4) designed and built by the National Research Laboratory (NRL). These were the first fully transistorized satellites. Vanguard was equipped with Western Electric GA-53233 and GF-45011 transistors, manufactured by the diffusion process. Explorer 1 was equipped with 33 transistors: four Western Union WE 53194 germanium diffused base transistors in the RF circuit; two germanium transistors (2N64); eight silicon 2N328 transistors manufactured by Raytheon; 19 silicon 2N335 grown junction transistors manufactured by TI. The Explorer 3 satellite, launched on March 26, 1958, featured 117 transistors, including 7 germanium made by Philco (L-5412), 4 made by Western Electric, 96 silicon made by TI (2N263 (switching circuit), TI 905 (grown junction) and TI 926); and 10 made by Raytheon (alloy junction) (2N328, 2N329, CK 791) (Ward 2007).

5.5. References

Bradley, W.E. (1953). The surface- barrier transistor. *Proceedings of the IRE*, 12, 1702–1706.

Burgess, P.D. (2008a). Transistor history. Semiconductor history: Faraday to Shockley [Online]. Available at: https://sites.google.com/site/transistorhistory/faraday-to-shockley.

Burgess, P.D. (2008b). Transistor history. RCA [Online]. Available at: https://sites.google.com/site/transistorhistory/Home/us-semiconductor-manufacturers/rca-history.

Burgess, P.D. (2010). Transistor history. Diffusion technologies at Bell Laboratories [Online]. Available at: https://sites.google.com/site/transistorhistory/Home/us-semiconductor-manufacturers/western-electric-main-page.

Burgess, P.D. (2011a). Transistor history. General Electric history – Semiconductor research and development at General Electric [Online]. Available at: https://sites.google.com/site/transistorhistory/Home/us-semiconductor-manufacturers/ general-electric-history.

Burgess, P.D. (2011b). Transistor history. Early semiconductor history of Texas Instruments. Grown diffused process [Online]. Available at: https://sites.google.com/site/transistorhistory/Home/us-semiconductor-manufacturers/ti.

Charpin, D.M., Fuller, C.S., Pearson, G. (1954). A new silicon p-n junction photocell for converting solar radiation into electric power. *Journal of Applied Physics*, 25, 676–677.

CHM (1951). The Silicon Engine Timeline. 1951: The first grown-junction transistor fabricated. Computer History Museum [Online]. Available at: https://www.computerhistory.org/siliconengine/first-grown-junction-transistors-fabricated/.

CHM (1952). The Silicon Engine Timeline. 1952: Transistorized consumer products appear. Computer History Museum [Online]. Available at: https://www.computerhistory.org/siliconengine/transistorized-consumer-products-appear/.

CHM (1954a). The Silicon Engine Timeline. 1954: Silicon transistors offer superior operating characteristics. Computer History Museum [Online]. Available at: https://www.computerhistory.org/siliconengine/silicon-transistors-offer-superior-operating-characteristics/.

CHM (1954b). The Silicon Engine, Timeline. 1954: Diffusion process developed for transistors. Computer History Museum [Online]. Available at: https://www.computerhistory.org/siliconengine/diffusion-process-developed-for-transistors/.

Christensen, H. and Teal, G. (1954). Method of fabrication of germanium bodies. Patent, US2692839.

Derick, L. and Frosch, C.J. (1957a). Manufacture of silicon devices. Patent, US2804405.

Derick, L. and Frosch, C.J. (1957b). Oxidation of semiconductive surfaces for controlled diffusion. Patent, US2802760.

Dezoteux, J. and Petit-Jean, R. (1980). *Les transistors.* PUF, Paris.

Felker, J.H. (1951). The transistor as a digital computer component. In *Proceedings AIEE-IRE Computer Conference.* Philadelphia, Association for Computing Machinery New York.

Fuller, C.S. (1962). Method of forming semiconductive bodies. Patent, US3015590.

Gardner, R.B. and Dill, F. (2010). The legendary IBM 1401 data processing system. *IEEE Solid State Circuits Magazine*, 2, 28–39.

Hall, R.N. (1955). P-N junction transistor. Patent, US2705767.

Hall, R.N. and Dunlap, W.C. (1950). P-N junctions prepared by impurity diffusion. *Physical Review*, 80(3), 467–468.

Haynes, J.R. and Shockley, W. (1949). Investigation of hole injection in transistor action. *Physical Review*, 75(4), 691.

Haynes, J.R. and Shockley, W. (1951). The mobility and life of injected holes and electrons in germanium. *Physical Review*, 87(5), 835.

Hoerni, J.A. (1962a). Method of manufacturing semiconductor devices. US Patent, 3025589. Semiconductor device. US Patent, 3064167.

Hoerni, J.A. (1962b). Selective control of electron and hole lifetimes in transistors. US Patent, 3184347.

Holonyak, N. (2007). The origins of diffused-silicon technology at Bell Labs, 1954-1955. *The Electrochemical Society Interface*, 30–34.

Hu, C.C. (2009). *Modern Semiconductor Devices for Integrated Circuits.* Prentice Hall, Upper Saddle River, NJ [Online]. Available at: https://www.chu.berkeley.edu.

Jackson, K.A. (2005). Processing of semiconductors. *Materials Science and Technology*, 21(1), 165.

Knight, J. (1957). The first RCA experimental, development and production transistors. *Tube Collector*, 10(4), 7–11.

Laws, D. (2007). The legacy of Fairchild in the Computer History Museum's visible storage exhibit. Working document, Computer History Museum, Mountain View.

Leamy, H.J. and Wernick, J.H. (1997). Semiconductor silicon: The extraordinary made ordinary. *MRS Bulletin*, 5, 47–55.

Lee, C.A. (1956). A high frequency diffused base germanium transistor. *Bell System Technical Journal*, 35, 23–34.

Lojek, B. (2007). *History of Semiconductor Engineering.* Springer, Berlin.

Lorenzini, P. (2021). Composants électroniques. Distribution des porteurs minoritaires dans transistor npn [Online]. Available at: http://users.polytech.unice.fr/Lorentz.

Moore, G.E. (1998). The role of Fairchild in silicon technology in the early days of Silicon Valley. *Proceedings of the IEEE*, 86(1), 53–62.

Morton, J.A. and Pietenpol, W.J. (1958). The technological impact of transistors. *Proceedings of the IRE*, 46, 955–959.

Ohl, R. (1952). Properties of ionic bombarded silicon. *Bell Technical Journal*, 31(1), 104–121.

Ohl, R. (1956). Semiconductor translating devices. Patent, US2750541.

Pankove, J. (1961). Transistors. Patent, US3005132.

Pearson, G.I. and Sawyer, B. (1952). Silicon p-n junction alloy diodes. *Proceedings IRE*, 40(11), 1348–1351.

Petriz, R.L. (1962). Contribution of materials technology to semiconductor devices. *Proceedings of the IRE*, 50(5), 1025–1038.

Riordan, M. and Hoddeson, L. (1997). *Crystal Fire: The Invention of the Transistor and the Birth of the Information Age*. W.W. Norton & Company, New York.

Saby, J. (1961). Broad area transistors. Patent, US2999195.

Scaff, J.H. and Theuerer, H.C. (1951). Semiconductor comprising silicon and method of making it. Patent, US2567970.

Schwartz, R. and Slade, B. (1957). A high-speed P-N-P alloy-diffused drift transistor for switching operations. *Electron Devices Meeting*, 3, 117.

Shive, J. (1949). The double surface transistor. *Physical Review*, 75, 689–690.

Shockley, W. (1948). Minority carriers and the first two transistors. *Bell Labs Lab Notebook*, 20455, 128–132.

Shockley, W. (1949). The theory of P-N junctions in semiconductors and P-N junction transistors. *Bell System Technical Journal*, 28(3), 435–489.

Shockley, W. (1951). Circuit element utilizing semiconductor materials. Patent, US2569347.

Shockley, W. (1957). Forming semiconductive devices by ionic bombardment. Patent, US2787564.

Shockley, W. (1976). The path to the junction transistor. *IEEE Transactions on Electron Devices*, 23(7), 597–620.

Shockley, W., Pearson, G.L., Haynes, J.R. (1949). Hole injection in germanium. *Bell System Technical Journal*, 28(3), 343–366.

Shockley, W., Sparks, M., Teal, G.K. (1951). P-N junction transistor. *Physical Review*, 83(7), 151–162.

Sparks, M.S. and Pietenpol, W.J. (1956). Diffusion in solids. *Bell Laboratories Record*, 12, 441 [Online]. Available at: www.americanradiohistory.com.

Sze, S.M. (2002). *Semiconductor Devices Physics and Technology*, 2nd edition. Wiley, New York.

Tanenbaum, M. and Thomas, D.E. (1956). Diffused emitter and base silicon transistors. *Bell System Technical Journal*, 35(1), 1–22.

Tanenbaum, M., Valdes, L.B., Buehler, E., Hannay, N.B. (1955). Silicon n-p-n grown junction transistors. *Journal of Applied Physics*, 26(9), 686–692.

Teal, G.K. (1955). Methods of producing semiconductive devices. Patent, US2727840.

Teal, G.K. (1976). Single crystals of germanium and silicon – Basic to the transistor and integrated circuit. *IEEE Transactions on Electron Devices*, 23(7), 621.

Teal, G.K. and Buehler, E. (1952). Growth of silicon single crystals and of single crystal silicon p-n junctions. *Physical Review*, 87, 190.

Theuerer, H.C., Kleimack, J.J., Loar, H.H., Christensen, H. (1960). Epitaxial diffused transistors. *Proceedings of the IRE*, 48(9), 1642–1644.

Ward, J. (2004). Historic transistor photo gallery. Transistor Museum [Online]. Available at: http://semiconductormuseum.com/photogallery.

Ward, J. (2005). Historic photo gallery. Transistor Museum [Online]. Available at: http://www.transistormuseum.com.

Ward, J. (2006). Historic photo gallery. Transistor Museum, TI GE 2N33X [Online]. Available at: http://www.transistormuseum.com.

Ward, J. (2007). The first transistors in space. A Transistor Museum interview with Dr. George Ludwig [Online]. Available at: http://www.transistormuseum.com.

Ward, J. (2016). Mike Warren – Transistor museum donation. Transistor Museum [Online]. Available at: http://www.transistormuseum.com.

The MOSFET Transistor

Research on the effect of an electric field on the conductivity of a doped semiconductor wafer, by William Shockley as early as 1945, led to the demonstration of a direct field effect as early as 1947 by John Bardeen and Walter (see Chapter 3, section 3.1.2.2).

But due to the problem of "surface states" (traps), it was not until 1959 that the first silicon MOSFET was produced by Martin Atalla and Dawon Kahng of Bell Labs.

It was the qualities of silicon oxide that led to the development of the silicon-based MOSFET transistors.

The first commercial transistors appeared in 1964.

The MOSFET transistor became the preferred component for computer logic circuits and memories in 1970, because of its low power consumption and miniaturization capacity.

Miniaturization by a factor of 1,000 has been achieved in 50 years, which means a reduction in the switching time by a factor of 1,000 and therefore an increase in operation speed.

The miniaturization of transistors has posed acute problems in terms of manufacturing processes, dielectric and electrode materials for the gate, source and drain, and electrical connections.

This chapter presents:

– the transistor operation (in static regime);

– the functions: switching and amplification;

– the CMOS component;

– the development history of the MOSFET transistor;

– the evolution of MOSFET transistor materials imposed by miniaturization (Moore's law).

6.1. Features and functions

6.1.1. *Introduction*

The field effect (presented in Chapter 3, section 3.1.1) consists of the polarization (conductivity inversion and/or modulation) of a thin surface layer of a wafer or thin film of a doped semiconductor, induced by the application of an electrostatic field perpendicular to the semiconductor surface via a flat electrode close to and parallel to the wafer surface.

Two types of field effects can be distinguished: depending on the direction of the transverse electric field applied (i.e. the polarity of the voltage applied to the electrode) and on the type (n or p) of semiconductor:

– direct field effect: modulation of the conductivity of a thin surface layer (majority carrier modulation) of a semiconductor wafer or thin film (the TFT transistor is presented in Volume 2, Chapter 3);

– reverse field effect, consisting of an inversion of the conductivity of a thin surface layer of a semiconductor P wafer, by creating an electron-enriched "inverted channel". This is achieved by applying a positive polarity to the "gate" electrode (see Figure 6.1(b)), or vice versa.

The first silicon inverse field-effect transistor MOSFET in silicon was produced by Martin Atalla and Dawon Kahng at Bell Labs in 1959.

6.1.2. *MOSFET transistor operation*

The current in a MOSFET transistor is a majority carrier current (electrons in an inversion channel N), whereas the current in a bipolar transistor is the sum of two currents, the main current being a minority carrier current.

Figure 6.1(a) shows the basic structure of a MOSFET transistor: on a P single-crystal substrate forming the base, 2 N islands (diffusion-doped n), the source and drain, and the depletion layers formed at the NP interfaces of the two diodes, plus a gate consisting of a dielectric layer (SiO$_2$) and a metal electrode (Al). Figure 6.1(b) shows the inversion channel formed when a positive voltage is applied to the gate.

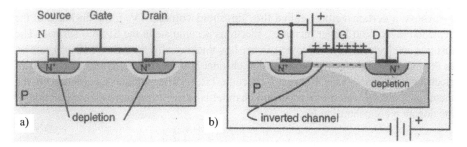

Figure 6.1. *MOSFET transistor: (a) structure and depletion layers; (b) under positive polarity of gate G, the "inverted channel" (Kuphaldt 2009, p. 7). For a color version of this figure, see www.iste.co.uk/vignes/silicon1.zip*

6.1.2.1. *MOS capacitor*

The central structure of the transistor is that of a metal-oxide-silicon (MOS) capacitor consisting of a P semiconductor substrate and a metal electrode separated by a dielectric layer (Figure 6.2). For a positive voltage V applied to the metal electrode, two equal and opposite charges Q develop on the surface of the two capacitor plates, with $Q = C_{ox} V$, where C_{ox} is the capacitance.

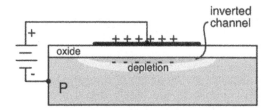

Figure 6.2. *MOS capacitor charged (Kuphaldt 2009, p. 7). For a color version of this figure, see www.iste.co.uk/vignes/silicon1.zip*

For a positive voltage at the metal electrode, the space charge on the metal electrode Q_m is positive. The holes in the P semiconductor are repelled, and a *depletion layer* is formed, which is made up of fixed negative charges Q_{dep} (dopant ions N_A) over a thickness Z, which increases with voltage[1]:

$$\text{For } V_G < V_{th} \quad Q_{dep} = e \ WL_GN_A \ Z \qquad\qquad [6.1]$$

---

1 All formulas quoted in this section are taken from Sze (2002, Chapter 6).

where N_A is the concentration of dopants (acceptors), L_G is the length and W the width of the gate.

Above a certain voltage called the "threshold voltage" (V_{th}), the thickness of the depletion layer Z no longer increases. Electrons accumulate at the Si/SiO$_2$ surface. The result is an electron-enriched layer of very low thickness (of the order of 10 nm), known as the "inversion channel". The inversion channel is formed by "thermal generation of electron–hole pairs" (see Chapter 1, section 1.2.3). The voltage V_{th} applied to the metal gate above which this inversion channel is formed is a quasi-linear function of the dopant p concentration N_A.

The charge of the inversion layer is:

$$Q_{inv} = WL_GC_{ox}. (V_{GS} - V_{th}) \qquad\qquad [6.2]$$

where W is the channel width, L_G its length and $C_{Ox} = \varepsilon_{Ox}/e_{ox}$, the grid capacity per unit area.

For a capacitor made of N semiconductor, the configurations are inverted.

6.1.2.2. MOSFET transistor operating modes

For a silicon P base MOSFET, called N-channel MOSFET (NMOSFET), the application of a positive gate voltage V_{GS} induces the injection of electrons from the S source, filling the holes in the depletion layer under the gate and, beyond the "threshold voltage" V_{th} , the formation of an electron-enriched layer known as the inverted channel, whose charge Q (formula [6.2]) can be modulated by the gate potential (Figure 6.1(b)).

Threshold voltage depends on the nature of the gate. For an aluminum gate, it is around 5 V. Modern transistors use highly doped n or p polysilicon, which allows much lower threshold voltages, of the order of 1 V (see section 6.2.2 and Figure 6.3).

Furthermore, the threshold voltage is a linear function of the dopant concentration p (N_A) in the substrate.

The characteristic $I_{DS} - V_{DS}$, current I_{DS} between source and drain as a function of the voltage between drain and source V_{DS}, with the grid-source voltage V_{GS} as parameter (Figure 6.3), comprise two regions corresponding to different regimes:

– in "linear" region, for $V_{DS} < V_{GS} - V_{th}$, the current between source and drain I_{DS} increases with the drain-source voltage V_{DS};

– under "saturated" conditions (plateau of the curves), $V_{DS} > V_{GS} - V_{th}$, the drain-source current I_{DS} is independent of the drain-source voltage V_{DS} and depends only on the voltage applied to the gate V_{GS}.

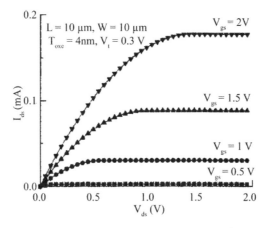

Figure 6.3. *Characteristic curves of a polycrystalline silicon gate MOSFET transistor with gate voltage as parameter (Hu 2009, p. 211)*

Figure 6.4 shows the inversion channel and the depletion layer in three regimes:

a) In the linear region where the inversion channel runs from source to drain, the current consists of electrons flowing in the inversion channel.

b) At the edge of saturation, for $V_{DS} = V_{Dsat} = V_{GS} - V_{th}$, the inversion channel only touches the drain at a pinch-off point.

c) Beyond saturation the inversion channel no longer joins the source and drain. Electrons arriving from the source via the shortened inversion channel are injected into the depletion zone of the drain and are accelerated toward the drain by the strong electric field applied by the drain-source potential V_{DS}.

In linear region, for a gate voltage $V_{GS} > V_{th}$, a continuous inversion channel is formed between the source and the drain of length L_G (Figure 6.4(a)), carrying a charge Q of electrons which increases with the potential applied to the gate ($V_{GS}-V_{th}$) (formula [6.2]). The electron current flowing between source and drain I_{DS} in the inversion channel is given by:

$$V_{DS} < (V_{GS} - V_{th}): I_{DS} = 2K \, (V_{GS} - V_{th})\} \cdot V_{DS} \text{ and } K = \mu_n \, C_{Ox} W / 2L_G \qquad [6.3]$$

where μ_n is the mobility of the majority charge carriers in the inversion channel (Chapter 1, section 1.2.7 and formula [1.10]).

In the saturated region, for a drain-source voltage $V_{DS} > V_{GS} - V_{th}$, the saturation current I_{DSat} flows between source and drain, independent of the voltage V_{DS} and depends only on the voltage applied to the gate V_{GS} for:

$$V_{DS} > V_{GS} - V_{th} \text{ and } I_{DSat} = K \ (V_{GS} - V_{th})^2 \text{ and } K = \mu_n \ C_{Ox}W/2L_G \qquad [6.4]$$

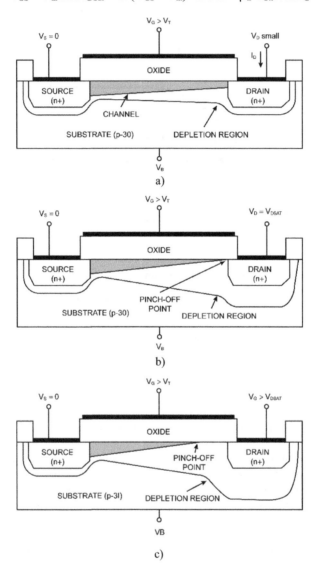

Figure 6.4. *The inversion channel (a) in linear region, (b) at the edge of saturation and (c) beyond saturation (Kang and Leblebici 2003, p. 65)*

The current between source and drain therefore varies as the square of the gate voltage and the transit time is equal to:

$$\tau_T = Q_{inv}/I_{DS} = L_G^2/\mu_n \cdot (V_{GS} - V_{th})$$
 [6.5]

Transit time is proportional to channel length, since reducing channel length reduces supply voltage by the same factor.

NOTE.– We will compare these expressions for the transit time in a MOSFET with those of the bipolar transistor in Chapter 5, section 5.1.2.2 and formula [5.4].

NOTE.– The mobility of majority charge carriers in the inversion channel μ_n is significantly lower than in a bulk semiconductor.

6.1.3. *Basic functions*

6.1.3.1. *Switching*

The MOSFET transistor is controlled as a component of an OFF/ON switching circuit and as a component of an amplification circuit, by the voltage applied to the gate V_{GS}.

The basic switching circuit is the inverter consisting of a MOSFET transistor and a load resistor R_D (Figure 6.5(a)):

$$V_{DS} = V_{DD} - R_{in} I_{DS}$$
 [6.6]

At saturation, the operating point of the circuit is determined by the intersection of the load line (formula [6.6]) with the plateau of the characteristic curve I_{DS}/V_{DS}, corresponding to the gate voltage V_{GS} (point Q in Figure 6.5(b)).

Switching is achieved by a voltage step V_{GS} applied to the gate, which results in a variation in current I_{DS} causing the transistor to switch from the OFF state (blocked, $I_{DS} = 0$) to the ON state, where current I_{DS} is defined by the point of intersection of the load line and the characteristic curve corresponding to the applied potential V_{GS} at the boundary between the linear (triode) and saturated regimes (Figure 6.5).

During the establishment of conduction, that is, for the duration of an OFF/ON switching operation, the electron current from the source gradually builds up the inversion channel the charge Q (formula [6.2]) of the MOS capacitor. The circuit operation can be modeled as a "parallel RC" source-grid circuit, where C is the MOS capacitance, $C = C$ WL_{oxG} (formula [6.2]) and R is the resistance of the gate-source circuit R_{in} through which the electrons forming the inversion channel are injected. This resistance can be deduced from formula [6.4], which relates the current I_{DS} to the gate voltage in saturation mode:

$$1/R_{in} = dI_{DS}/dV_{GS} = (W/L)\mu_n C_{ox}(V_{GS} - V_{th}) \tag{6.7}$$

Hence, the circuit time constant RC:

$$\tau_{GS} = R_{in} \cdot C_{ox} W L_{inv} = L^2/\mu_n(V_{GS} - V_{th}) = \tau_T \tag{6.8}$$

The time constant of the τ_{GS} circuit is equal to the transit time τ_T of electrons at saturation.

The switching time is the storage time OFF/ON, τ_C, time required to charge the MOS capacitance (Figure 6.1(b)) following a gate voltage "jump" V_{GS}. It is therefore equal to three times the transit time of electrons from source to drain (formula [6.8]). The same result applies as for the bipolar transistor (formula [5.14]):

$$\tau_C = 3\,\tau_T \tag{6.9}$$

In 1988, for a channel length of 1 μm, the switching time reached 100–200 ps. An IBM team has developed a transistor with a channel length of 100 nm, with a switching time of 20 ps (10^{-12}) (Meindl 1977; Bois and Rosencher 1988).

In 2007, for $L_G = 25$ nm, the switching time was 5 ps.

6.1.3.2. Amplification

A weak signal is amplified by a transistor biased in the saturation region. The operating point Q is defined by the intersection of the load line imposed by the inverter with the step in the characteristic curve corresponding to the voltage applied to the gate V_{GS} (Figure 6.5(b)).

At low frequencies (weak signal applied to the gate), no current flows in the grid-source circuit. All of the current (of electrons) from the emitter flows through the base. The transistor is a resistor R_{in} (formula [6.7]). The voltage gain of the signal to be amplified $v_{GS}(t)$ is equal to (in A of Figure 6.5(b)):

$$v_{ds}(t)/v_{GS}(t) = -R_D/R_{in} \tag{6.10}$$

At high frequencies (Hu 2009, p. 230), for a sinusoidal signal applied to the grid:

$$v_{GS}(t) = a\sin\omega_T t \tag{t[6.11a]}$$

A fraction of the current $i_{GS}(t)$ from the emitter flows through the gate-source circuit via the gate capacitance C_G:

$$i_{GS}(t) = j\,C_G\,\omega_T \cdot v_{GS}(t) \tag{6.11b}$$

and a fraction of the current $i_{DS}(t)$ flows through the base, as a function of the signal $v_{GS}(t)$ (formula [6.7]) (B (i_d) in Figure 6.5(b)):

$$i_{DS}(t) = (1/R_{in}) \cdot v_{GS}(t) \tag{6.12}$$

and the current gain is:

$$i_{DS}(t)/i_{GS}(t) = 1/j \ R_{in} \cdot C_G \ \omega_T \tag{6.13}$$

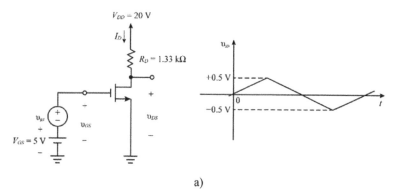

a)

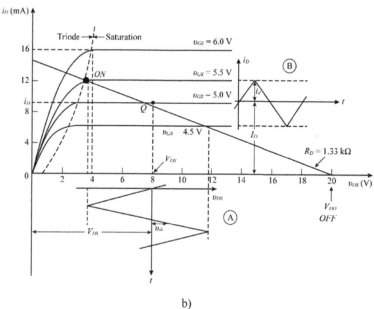

b)

Figure 6.5. *(a) Inverter circuit and signal to be amplified; (b) graphical representation of switching (ON/OFF) and amplification (A and B) functions performed by an inverter (Redoutey (n.d.))*

The cut-off frequency is defined as the frequency above which all of the amplified current i_{DS} passes through the grid-source circuit, that is, when the ratio of the two currents is equal to 1:

$$f_T = 1/2\pi\ \omega_T = 1/2\pi\ R_{in}\cdot C = \mu_n\cdot(V_{GS} - V_{th})/2\pi L^2 = 1/2\pi\ \tau_T \qquad [6.14]$$

where τ_T is the electron transit time (formula [6.5]).

The higher the cut-off frequency f is, the shorter the channel length. In 2007, for $L_{node} = 45$ nm, that is, $L_G = 25$ nm and $f_T = 200$ GHz.

6.1.4. *The CMOS component*

The CMOS (complementary MOS) component, invented by Frank Wanlass of Fairchild in 1963, consists of two coupled MOSFET transistors of opposite polarity (Figure 6.6). The gates are connected to the same input, and the two drains are connected to the same output. This is an inverter (Wanlass and Sah 1963; Moore 1964; Wanlass 1967). The advantage of this component is that it eliminates the resistance in the classic inverter structure.

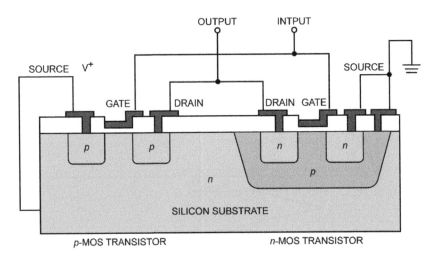

Figure 6.6. *CMOS component structure (Meindl 1977)*

As the two transistors have opposite PNP and NPN polarity, when one conducts, the other is blocked, and vice versa. The complementary silicon MOS with planar structure operates with very low power consumption. It was this low power consumption that established CMOS as the key component in logic circuits and SRAM memories.

6.1.5. *MOSFET transistor development history*[2]

The hypothesis of the formation of an n-type inversion channel in a block of P semiconductor by applying an electric field was formulated by John Bardeen in December 1947 to explain the result of an experiment described in Chapter 3, section 3.1.2.2 ("that an inversion layer was being induced electrically by the strong field under the droplet") (Bardeen 1956).

Research into the "field effect" carried out by Bell Labs at the end of the Second World War under the impetus of William Shockley did not come to fruition until a decade later, due to the presence of "surface states as traps" which prevent the flow of electrons in a thin layer on the surface of a P semiconductor plate or film (Chapter 3, section 3.1.2.1).

An intensive program on the surface trap problem was launched. John Bardeen worked with physicists Walter Brattain and Gerald Pearson. But for about 10 years, until 1958, the presence of these traps prevented Bell Labs researchers from building the MOSFET transistor. It was necessary to obtain a surface with a very low density of these surface traps (of the order of $10^{15}/m^2$), but nobody knew how to achieve such a surface condition.

It was while pursuing studies on the bipolar transistor that Bell Labs researchers discovered the existence of an "inversion channel". This observation led directly to the MOSFET transistor.

Researchers at Bell Labs, continuing studies on the germanium-based bipolar transistor, had observed that the flow of minority carriers (electrons), injected into the P-base between emitter and collector, was much greater than expected. It was postulated that a channel existed between the two junctions N-P and P-N, bridging the gap between emitter and collector. Since the emitter and collector were of type n, these two regions were connected by an "inversion channel N". In 1953, Brown modeled the behavior of such a channel and experimentally verified its conductivity. Replicating the experimental set-up used by Bardeen and Brattain (Chapter 3, Figure 3.7), a block of germanium N block, one surface of which was coated with an electrolyte

2 For a full review, see the article by one of the authors of the discovery (Sah 1988).

(glycol diborate) that formed a P layer, Brown showed that the electrolyte ions absorbed at the surface of the P layer repelled the majority charge carriers of the P layer, namely holes, and attracted minority charge carriers, namely electrons, thus creating an inversion channel at the P-layer surface (Brown 1953).

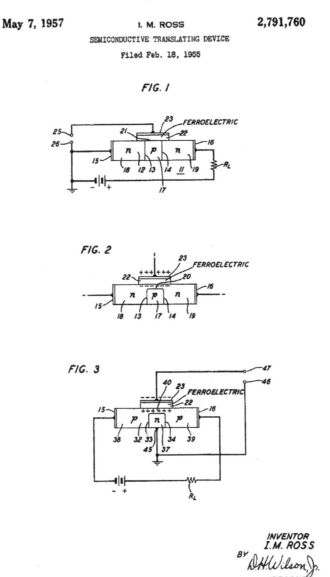

Figure 6.7. *Germanium field-effect transistor (Ross 1957)*

The design and manufacture of the first inverse field-effect transistor are thanks to Ian Ross in 1955 (Ross 1957), showing that such an "inversion channel" could be obtained by placing a gate consisting of a metal electrode and a dielectric on the P base of an N-P-N bipolar transistor (Figure 6.7). When a positive voltage is applied to the electrode, an N surface layer is formed on the surface of the P base, allowing the flow of electrons between the transistor's emitter and receiver. The base material was germanium, and the dielectric insulating the gate from the base was a ferroelectric material. This patent was described by the inventors of the MOSFET transistor as follows: "Subsequently, a monumental proposal was made by Ian Ross" (Kahng 1976). However, deposition of the dielectric layer proved difficult to achieve industrially, and it does not appear that attempts were made to produce a germanium MOSFET.

The development of the silicon MOSFET transistor was prompted by a suggestion from Martin Atalla at Bell Labs, namely, that the dielectric layer insulating the metal electrode from the base could be produced by a "very stable" oxide layer formed at high temperature on the surface of a silicon single crystal. This suggestion followed the discovery in 1958 by Atalla's team of the formation, temperature of 920°C for 10–30 min in dry oxygen of an oxide layer (15–30 nm) (Atalla 1965). This layer eliminated most dangling bonds (Atalla et al. 1959) (see Chapter 3, section 3.1.2.1, Figure 3.5) and absorbed impurities present in the crystal near the surface by diffusion: the gettering effect is described in Chapter 2, section 2.3.1.3. While the surface traps were neutralized, an external electric field could penetrate the surface of the silicon substrate, creating this inversion layer (excess electrons), thereby increasing conductivity and thus the electron current between source and drain.

With the traps neutralized, Kahng and Atalla, the inventors of the MOSFET transistors, produced a practical field-effect transistor in 1960 (Figure 6.8), whose structure is identical to that shown in Figure 6.1 (12: silicon base N (p-doped) 6 Ω·cm; 13–14: source and drain (p-doped); 15: inversion channel (50 µm); 16–17: PN junctions; 19: oxide layer (100 nm thick); 21: Al gate electrode, much wider than the channel). The oxide layer and aluminum electrode are deposited after the source and drain have been made by diffusion (Kahng and Atalla 1960).

This discovery stems from studies undertaken by the group led by Martin Atalla at Bell Labs, back in 1955, which led to the realization of bipolar transistors with planar configuration (see Chapter 5, section 5.3.3). These studies were undertaken to solve the problem of surface deterioration of silicon wafers during high-temperature diffusion doping operations and led to the protection of the wafers by the formation of a layer of amorphous silica oxide (see Chapter 5, section 5.3.1, oxide masking).

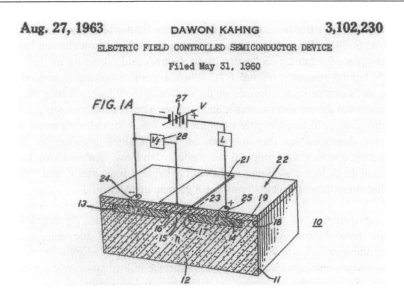

Aug. 27, 1963   DAWON KAHNG   3,102,230

ELECTRIC FIELD CONTROLLED SEMICONDUCTOR DEVICE

Filed May 31, 1960

FIG. 1A

Figure 6.8. *The PMOSFET (p channel) transistor (Kahng 1963)*

Only the silicon MOSFET transistor could be realized. The formation of silica in situ eliminated the surface traps, which prevented the field effect. The silica formed in situ also constituted the gate's dielectric layer. Neither germanium nor the GaAs semiconductor can be used, as good insulators cannot be produced on their surfaces, making them unsuitable for the manufacture of field-effect transistors.

The first silicon MOSFET transistor was produced in 1960 by Bell Labs researchers. But Bell Labs did not pursue studies on these transistors, of little interest (according to them) for their telephone system circuits. Dawon Kahng designed and built the "floating gate" transistor in 1967 (Volume 2, Chapter 2, section 2.3).

By contrast, in the same year, two companies, Fairchild (Moore 1998) and RCA, were pioneers in the development of MOSFET transistors. In June 1960, as soon as Bell Labs researchers Kahng and Atalla presented their MOSFET transistor at a conference in Pittsburgh, PA (Kahng 1960), Chih-Tang Sah (Sah 1961), from Fairchild, and Karl Zaininger and Charles Mueller, from RCA, manufactured a MOS transistor in 1960 (CHM 1960).

The MOS transistor was developed at RCA by Fred Heiman and Steven Hofstein in 1962 (Hofstein and Heiman 1963).

The first three commercial MOSFET transistors were announced at the end of 1964 by three companies: General Microelectronics (GME 1004), Fairchild (F100-PMOSFET) and RCA (3N98-NMOSFET) (CHM 1960). General Microelectronics was founded in 1963 by defectors from Fairchild. When Frank Wanlass left Fairchild and joined GME in 1964, the company concentrated on MOSFET transistors.

In 1962–1963, research at Fairchild led to a major innovation: the CMOS (Complementary MOS device) by Frank Wanlass (Figure 6.6) (Wanlass and Sah 1963). Frank Wanlass was recruited by Fairchild in August 1962.

But skepticism about the value of this component was widespread, even within the Fairchild company. Frank Wanlass left Fairchild to join General Micro-electronics (GME) in 1964 (Lojek 2007, p. 334).

In 1968, there were only two vendors of CMOS-based components, including RCA and a small company, Solid State Scientific Devices.

6.1.6. *The effect of impurities on the dispersion and instability of the electrical characteristics of transistors*

A wide dispersion and instability of the electrical characteristics of silicon MOSFET transistors were observed in production, due to intrinsic material problems: "traps" at the silicon/oxide gate interface (presented in Chapter 3 section 3.1.2, Chapter 2 section 2.3.1.3, and Chapter 5 section 5.3.1) and contamination during manufacturing operations. The companies involved devoted all of their resources to solving these problems.

Studies aimed at reducing "traps" at the silicon/silicon oxide interface were continued. In 1965, Balk (1965) (IBM) showed that by replacing thermal oxidation by oxidation in the presence of water vapor, trap blocking was achieved through the formation of hydrogen bonds. Trap blocking is achieved industrially by heat treatment at 450°C in a hydrogen atmosphere. Also in 1965, it was shown by several experimenters, including Balk, that the density of "traps" was significantly lower on oxidized silicon surfaces with {100} orientation than on {110} and {111} surfaces. The result of these treatments was to reduce the "trap" density by a value observed by Shockley and Pearson, in 1948, of 10^{13} traps/cm^2 to a value 10,000 times lower, of the order of 10^9 traps/cm^2 (Balk et al. 1965).

The source of instability in the electrical characteristics of silicon transistors (unstable threshold voltage V_{th}) was identified by Fairchild researchers in 1965 (Snow et al. 1965): sodium contamination of the gate dielectric. The Na$^+$ ions present on the dielectric surface (during layer formation) are mobile and migrate into the oxide under the influence of the electric field applied to the gate. These to-and-fro movements in the oxide, which varied the MOS capacitance, caused problems with the transistor's threshold voltage stability. The source of this contamination was the tungsten filament constituting the heating element, causing the vaporization of the aluminum deposited on the transistor's gate. Solutions were found to eliminate sources of contamination during manufacturing operations.

Solutions based on the addition of layers of materials acting as getter were developed:

– the formation of a layer of PSG (phosphosilicate glass) on the surface of the oxide layer, acting as a getter for Na^+ ions (gettering effect), by Kerr and Logan (1964) of IBM;

– the production of a gate insulator consisting of a 60 nm layer of silica and a 40 nm layer of silicon nitride (Si_3N_4), forming a barrier to Na^+ or K^+ ions deposited during manufacturing operations, by Sarace of Bell Labs (Sarace et al. 1968) (see section 6.2).

Iron contamination (see Chapter 4, section 4.1.4.2) was also observed. It significantly reduces the breakdown voltage of the gate oxide for dielectric thicknesses greater than 5–10 nm, through the formation of Fe-Si or Fe-Si$_2$ precipitates at the Si/SiO_2 interface, which penetrate the oxide, increasing the electric field at the tip of the precipitate in the oxide, hence, the need to eliminate them. The value of 2×10^{11} atoms/cm^3 for the concentration of iron dissolved in silicon can be considered as a limit below which the integrity of the gate oxide layer (gate oxide integrity (GOI)) is not affected for oxide layer thicknesses greater than 5 nm (Istratov et al. 2000).

Requirements for iron concentrations in silicon wafers increased from 10^{11} atoms/cm^3 in 1995 (0.002 ppba) to 10^{10} atoms/cm^3 in 2004 and reached 5×10^9 atoms/cm^3 in 2007 for integrated circuits (SIA 1999). Concentrations reached in CZ single crystals are below 10^{10} atoms/cm^3.

These metallic impurities, iron in particular, are introduced during manufacturing operations by contact with equipment, and penetrate and diffuse into the material during heat treatment in certain manufacturing stages. As a result, performance is degraded by contamination.

Studies carried out by numerous companies, in particular Fairchild, then Intel, IBM and many others, mainly in the United States, ensured the future of the silicon MOS transistor and prompted Bell Labs to resume its studies. This led to the invention of new MOS structure components (Volume 2, Chapter 2).

6.2. MOSFET miniaturization and materials

6.2.1. *Miniaturization: Moore's law*

According to Berry (2017, p. 88):

> The dominant phenomenon in the circuit industry has been Moore's Law stipulated in 1975 by Gordon Moore, creator of Intel. According to this law, the number of transistors per unit area in integrated circuits would roughly double every two years. This law has been perfectly respected ever since. But the word "law" is a little misleading here, as it's not a law at all in the sense of physics: it's nothing more than a concerted economic decision by the entire industry to meet this deadline. Its extraordinary success is clearly to the credit of semiconductor materials physicists, who have made technological advances requiring enormous imagination and skill to push back all obstacles.

It is with MOSFET transistors that this law has been followed.

One of the essential features of MOSFET transistors is their miniaturization, enabling the quest for ever greater computing power while minimizing the energy consumed by each switch. The switching time is proportional to the transit time from source to drain, which is proportional to the length of the inversion channel (formula [6.9]).

If we reduce all of the dimensions of the transistor by a factor k (miniaturization), in particular the length (or rather the surface area (LW)) of the inversion channel and the thickness of the dielectric layer e_{ox}, to maintain the same electric field $E = V_{GS}/e_{ox}$, required to create the inversion channel, the voltage applied to the gate electrode V_{GS} must be reduced by the same factor. The switching time is reduced by the same factor k and the power dissipated by switching is reduced by a factor k^2:

$$W = 1/2 \ Q \ V_G = 1/2 \ C_G \ V_{GS}^2$$

The density of transistors on the integrated circuit increases by a factor of $1/k^2$, while the power dissipated per unit area of the chip remains unchanged. The result is an integrated circuit with more elements, switching faster and consuming less power; greater integration therefore leads to faster logic.

Table 6.1 shows the variations in the geometric parameters characterizing the MOSFET transistor, based on major developments. Note that we have gone from a channel length of 25 μm in 1960 to a channel length of 25 nm in 2015, that is, miniaturization by a factor of 1,000.

The miniaturization of transistors posed acute problems in terms of manufacturing processes, source and drain materials, dielectrics and gate electrodes, and source and drain electrical connections. Hence, the search for new solutions, each of which took between 10 and 13 years to develop, which testifies to the problems encountered.

6.2.2. *Gate materials*

6.2.2.1. *Polycrystalline silicon gate electrode (self-aligned gate) and new manufacturing process*

This "technology" (silicon-gate technology), developed between 1963 and 1967 (CHM 1968) by Bell Labs researchers, enabled the development and semi-miniaturization of the MOSFET transistor and integrated circuits.

The high scrap rates observed in manufacturing until 1965, the much lower switching speed than that of bipolar transistors manufactured with the same planar configuration, the high threshold voltage to be applied to the aluminum gate (5 V), implying very high transistor operating voltages (25 V), preventing a certain degree of compatibility with the operation of a bipolar transistor on the same chip, led several companies from 1965 onwards to launch R&D programs aimed at remedying these deficiencies and shortcomings.

Until now, the manufacturing process of an NMOSFET transistor consisted, in the first step, of producing the source and drain regions, delimited by etching in the P base and doped by the CVD process (Chapter 5, section 5.2.4.1), followed by the gate: the gate dielectric by oxidation, the gate electrode by low-temperature deposition of an aluminum layer. This process avoided a reaction of the aluminum with the underlying silica. The difficulty of aligning (positioning) the photolithographic masks delimiting source and drain regions, then the gate (Volume 2, Chapter 1, section 1.4.3.3), very high tolerances on gate length with an overlap between the gate and source and drain regions, of the order of 8–10 μm, were necessary to ensure transistor operation, thus minimizing scrap rates, but at the expense of switching speed and electrical power consumption.

The manufacturing process developed by Kerwin et al. (Figure 6.9(a)) involves a reversal of the original manufacturing process (Kerwin et al. 1969): the gate (dielectric and electrode) is made first (step 7), then the "source and drain" are made by high-temperature diffusion doping (step 10) (or, more recently, by ion implantation). "This modification to the manufacturing process necessitated a change of material for the gate electrode". During the diffusion heat treatment, an aluminum gate electrode would have reacted with the silica gate dielectric. Polycrystalline silicon (doped by the subsequent diffusion operation and thus becoming conductive) was the solution chosen for the gate material. One of the advantages is to have an extremely "stable" poly-Si/silica contact, minimizing surface states. This solution made it possible to produce transistors whose dimensions (gate and channel length) and gate/source or drain overlaps were perfectly controlled and, above all, amenable to advanced miniaturization. Gate/source or drain overlap areas are of the order of 1 nm.

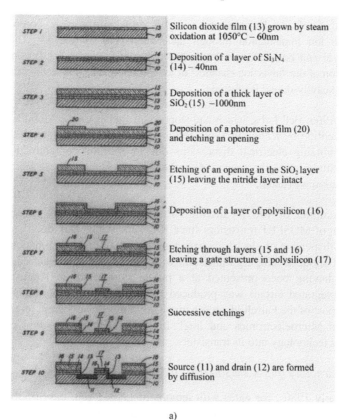

a)

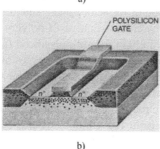

b)

Figure 6.9. *MOSFET transistor with silicon gate electrode and source and drain achieved by diffusion or ion implantation: (a) sequence of manufacturing operations (Kerwin et al. 1969); (b) achieved structure (Meindl 1977)*

The sequence of manufacturing operations developed by Kerwin et al. is shown in Figure 6.9(a).

In subsequent developments, deposition of an epitaxial layer of undoped silicon on the P silicon substrate, is made prior to deposition of layers 13 and 14, then, between steps 5 and 6 of the Kerwin process, this epitaxial layer is boron-doped by subcutaneous ion implantation through dielectric layers 13 and 14. The undoped epitaxial layer can be locally ion-implanted (Chapter 5, section 5.3.4) using masks to create a region of the "inverting channel" of low p-doping, and thus control the threshold voltage on a heavily-doped P substrate.

In early 1966, Bell Labs, which had stopped all studies on the MOSFET transistor since its invention by Kahng in 1960, resumed studies, apparently with the primary aim of minimizing electrical dispersions due to Na^+ ion contamination (section 6.1.6). The study led to the process described above (Figure 6.9(a)). But here again, no further action was taken.

In the summer of 1967, Fairchild, which had been heavily involved in the development of MOSFET transistors since 1963, and which had identified the cause of the wide dispersion of electrical characteristics – Na^+ ion contamination – as early as 1964, set about developing transistors with polysilicon gate electrodes (the Bell Labs work having been presented at a public symposium in August 1967). A prototype integrated circuit was produced by Fairchild in 1968: the 3708 (CHM 1968). But most of the Fairchild people involved in this development left Fairchild to join General Microelectronics and Intel. In 1969, Intel was the first company to integrate this technology into its transistors.

6.2.2.2. Refractory metal silicide gate electrode

In the early 1980s, for gates with lengths of $L_G < 2$ μm, the Si-polycrystalline gate electrode was no longer suitable, as it offered too high a resistance (insufficient doping). Gate electrodes made of refractory metal silicide $TiSi_2$, $TaSi_2$, WSi_2, with a resistivity 10 to 50 times lower than that of Si-polycrystalline, were produced. In fact, a sandwich consisting of a thin film of polysilicon (poly-Si) (thickness 0.2 μm) and a film of silicide was made to retain the advantages of very good adhesion of the base silicon to the SiO_2 gate insulator, making: a "stable" interface (Figure 6.10). This silicide film eliminates another phenomenon affecting the interconnection between the poly-Si gate electrode and the aluminum electrical contact (Figure 6.11): electromigration of silicon from the grain boundaries of the poly-Si layer to the aluminum, leading to dissolution of the silicon layer. The same problem arises for interconnections between the source or drain and the aluminum electrical contact. The silicide film acts both as a barrier and as an ohmic contact layer.

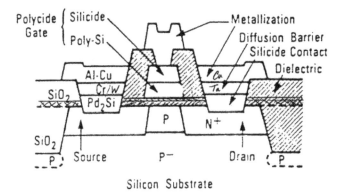

Figure 6.10. *Cross-section of a MOSFET with Ti, Ta or W silicide gate electrode and SiON gate dielectric (Sequeda 1985)*

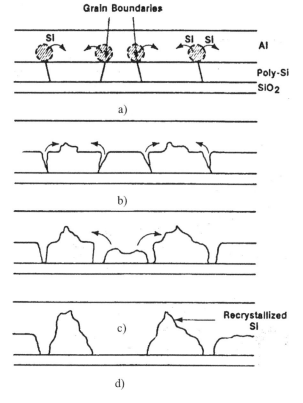

Figure 6.11. *Electromigration of silicon from the poly-Si gate into the aluminum layer, causing the gate to disappear (Pramanick and Saxena 1983)*

6.2.2.3. *Silica dielectric gate*

Miniaturization leads to a reduction in the thickness of the insulating (dielectric) layer of the gate, and the voltage applied to the V_G gate electrode is reduced by the same factor to obtain the same electric field:

$$E = V_{GS}/e_{dielectric} \text{ with } Q/S = E.\ \kappa$$

With a silica dielectric SiO_2, $\kappa = 3.9$, the electric field required to create a "channel" of sufficient conductivity of 0.1mmho is of the order of 10^6 V/cm (1 V/10 nm).

Initially, in 1966, with the switch to the polycrystalline silicon gate electrode the dielectric layer consisted of a 60 nm thick SiO_2 layer and a 40 nm thick Si_3N_4 layer (Sarace et al. 1968). In 1976, the thickness of the dielectric layer was reduced to 70 nm, in the early 1980s to 20 nm and in 1997 to 5 nm (12 atomic layers). In 2001, with the Intel Pentium 4, the dielectric layer thickness reached 2 nm.

For dielectrics with a thickness of less than 2 nm, diffusion of boron, from the polysilicon gate electrode (heavily doped), through the oxide layer enriches the channel material and modifies the properties of the "gate", in particular the threshold voltage. Incorporating nitrogen into the dielectric reduces boron diffusion.

A practical limit for the silica layer thickness is of the order of 1.2 nm (5 atomic layers), albeit with significant leakage currents through the insulator, of 1–10 A/cm^2, but still allowing acceptable transistor operation. Transistors for very low-power applications require leakage currents of less than 10^{-3} A/cm^2.

6.2.2.4. *Silicon oxynitride gate dielectric*

Replacing the dielectric SiO_2 (dielectric constant $\kappa = 3.9$) with a high dielectric constant dielectric (Si_3N_4: $\kappa = 7$) allows the increase of the thickness of the dielectric layer. The field strength is reduced, but the charge density in the inversion channel is maintained (maintaining the same ratio κ/ε) and therefore conductivity, while reducing the leakage current through the insulator. In the 1990s, the industry produced a gate dielectric in silicon oxinitride SiO_xN_y, with a higher dielectric constant, by thermal oxidation of silicon in the presence of nitrogen, while maintaining a very stable interface and a very low surface trap density. In 2002, using a 1.5 nm thick SiO_xN_y dielectric, leakage currents of the order of 10^{-4} A/cm^2 were obtained.

6.2.3. *New HKMG gate and new manufacturing process*

6.2.3.1. *New HKMG gate with high dielectric constant and refractory metal gate electrode*

The work leading up to the new HKMG (high-k gate dielectrics/metal gate) was completed in 2007 (Wilk et al. 2001; Bohr 2007).

As early as 1987, for a gate length of < 1 μm, it was decided to replace the refractory metal silicide gate electrode with a refractory metal electrode: tungsten W; Mo offering even lower electrical resistance.

Furthermore, for dielectric thicknesses below 1.2 nm, it became imperative to replace the SiO_xN_y dielectric.

The oxides TiO_2, ZrO_2, HfO_2 of the same valence and structure, with very high dielectric constants (κ 10–80), were to be suitable materials to replace SiO_2. TiO_2, although it has a high dielectric constant, 80–110, was not chosen as a dielectric material due to the existence of sub-oxides. Such sub-oxides present oxygen vacancies, which act as traps for charge carriers.

Zirconium and hafnium oxides were selected. But these dielectric layers had to be produced using a deposition process. However, the CVD deposition process, which requires a high temperature, does not produce a "stable" interface free of surface traps (dangling bonds) with silicon.

6.2.3.2. *The ALD deposition process*

The problem was solved using the ALD (atomic layer deposition) process: deposition of successive monomolecular layers (alternating flux deposition) (Wilk et al. 2001; George 2010; Johnson et al. 2014). This process was "invented" in 1974 by T. Suntola (George 2010) to enhance the quality of ZnS films for electroluminescent displays.

This process is derived from the CVD process, where the chemical reaction between two gaseous reactants takes place on the surface of the deposit being formed, at high temperature:

$$ZrCl_4(g) + 2H_2O(g) \rightarrow ZrO_2(s) + 4ClH(g)$$

The ALD process consists of carrying out this reaction in two stages, consisting of two elementary reactions, which enable the structure of the silicon to be reproduced identically, monomolecular layer after monomolecular layer, allowing strict control of the structure and thickness of the dielectric layer. To avoid any reaction between the oxide and the substrate, the deposition temperature must be much lower than that of CVD.

The first step involves creating a layer of SiO_2 oxide on the surface (100) of the silicon substrate, saturated by chemisorption of O-H bonds through thermic oxidation (Figure 6.12(a)). With the surface bonds saturated, the reaction stops on its own, the residual gas phase is eliminated and a second reaction is initiated by introducing the second reagent, $ZrCl_4$, into the reactor, the basic reaction being the substitution of the two OH groups by $ZrCl_2$ groups (Figure 6.12(b)). Then, by introducing H_2O, the first monomolecular layer of ZrO_2 is formed (Figure 6.12(c)). Once the excess reagents and ClH reaction products have been eliminated, the process can be restarted, and a second ZrO_2 layer will be created, and so on.

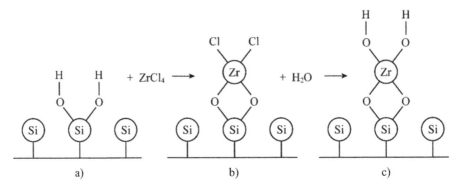

Figure 6.12. ALD process for forming a SiO_2 layer followed by formation of a ZrO_2 layer on a single-crystal silicon substrate

With ZrO_2 or HfO_2, using the ALD process (with $ZrCl_4$ as precursor), a uniform (conformal) layer of around 2.5-nm thickness is obtained on a 1.5 nm SiO_2 layer, forming a very high-quality interface (Figure 6.13).

In addition, the oxides ZrO_2 and HfO_2 ($\kappa = 20$) form stable silicates with silicon. Hence, the development of dielectrics using SiO_2-ZrO_2 (or HfO_2) "alloys", which combine the properties of both materials to form a two-phase system: a polycrystalline material with a high dielectric constant and an amorphous SiO_2 material. The addition of SiO_2 creates a "stable" amorphous film on silicon (with very low leakage currents).

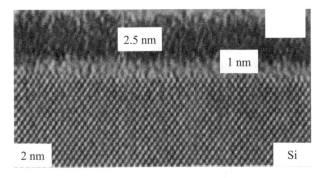

Figure 6.13. *TEM image of an HfO₂ oxide film (2.5 nm) deposited by the ALD process on a layer of SiO₂ (1 nm) on a Si wafer (Gusev et al. 2003)*

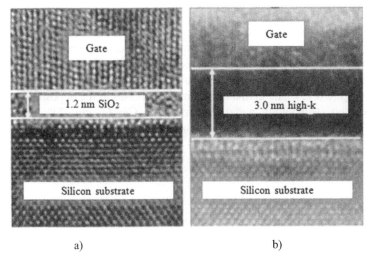

a)   b)

Figure 6.14. *TEM images: (a) 90 nm/65 nm generation MOSFET gate (poly-Si gate (1.2 nm SiO₂)); (b) 45 nm generation gate (3.0 nm high-k). EOT (Intel 2006)*

This process, derived from the CVD process, requires relatively low deposition temperatures, so it was necessary to return to the original manufacturing process, with the gate being formed after high-temperature doping of the source and drain zones. It should be noted that this dielectric requires a layer of SiO₂ (of the order of 0.5 nm) to ensure surface "stability" (minimization of surface traps).

The poor performance of components made with a hafnium silicate dielectric gate and a polycrystalline silicon electrode made it necessary to change the gate electrode material from refractory metal silicide to a refractory metal, W, Mo, with even lower electrical resistance, but with the same "work function" as silicon (see Chapter 1 section 1.2.4.1 and Figure 1.7). The electric field created is a function of the difference between the work function of the silicon in the channel and that of the gate metal.

This new HKMG (high-k gate dielectrics/metal gate) grid was introduced in 2007 by Intel for its 45 nm generation (technology node) components (Figure 6.14) (Wilk et al. 2001; Bohr et al. 2007).

Leakage currents through the HKMG gate are reduced by a factor of 25 for NMOS transistors and by a factor of 1,000 for PMOS transistors compared with a SiON/poly-Si gate. Switching times are thus faster.

6.2.4. Source and drain materials (strained silicon)

The distortion of silicon's crystal lattice alters the electrical properties of the channel, increasing electron and hole mobility of electrons and holes. In PMOS (P-channel MOSFET) transistors, this is achieved by doping the silicon with germanium in the source and drain regions, resulting in uniaxial compression in the channel (Figure 6.15). The germanium content increases in the opposite direction of the generation (technology node) (90 nm, 17% Ge; 65 nm, 23% Ge; 45 nm, 30% Ge), while the length of the inversion channel (electron/hole path from source to drain) remains constant at around 25–24 nm.

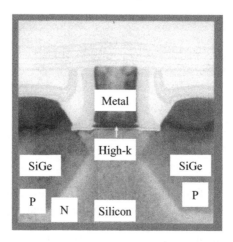

Figure 6.15. *MOSFET P-channel transistor (third generation source and drain: Si-30% Ge (see Table 6.1) (Mistry 2007)*

For NMOS (N-channel MOSFET) transistors, this is achieved by doping the source and drain with carbon at 0.25%.

6.2.5. *New architectures*

These new architectures are presented for information only. The aim of these architectures is to confine (channel) the electron current from source to drain to limit leakage from the inversion channel (Figure 6.16(a)) by reducing the space available to the electron current, without reducing the length of the channel (electron/hole path from source to drain).

6.2.5.1. *Fully depleted silicon on insulator MOSFET transistor*

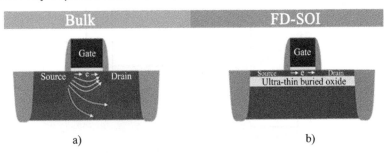

a)   b)

Figure 6.16. *(a) MOSFET transistor; (b) FD-SOI transistor MOSFET (STMicroelectronics: the FD-SOI innovation). For a color version of this figure, see www.iste.co.uk/vignes/silicon1.zip*

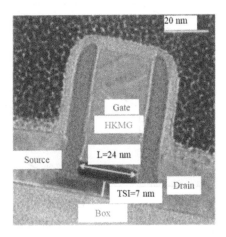

Figure 6.17. *FD-SOI transistor (Fenouillet-Beranger 2014). For a color version of this figure, see www.iste.co.uk/vignes/silicon1.zip*

An ultra-thin layer of buried oxide insulator is "deposited" on the surface of the silicon base. A layer of (monocrystalline) silicon is deposited on top of the oxide layer. This silicon layer, inserted between the two oxide layers, forms the channel through which electrons are injected by the source and captured by the drain flow.

6.2.5.2. *FinFET MOSFET transistor*

In 2011, Intel replaced the planar structure with a 3D FinFET Tri-Gate structure (Figures 6.18 and 6.19) developed by the University of California at Berkeley. The volume of the inversion channel is limited by the three faces of the dielectric and the gate. The length of the inversion channel is always the same as in the conventional structure. The transit time is therefore greatly reduced (Liu 2012; Stevenson 2013).

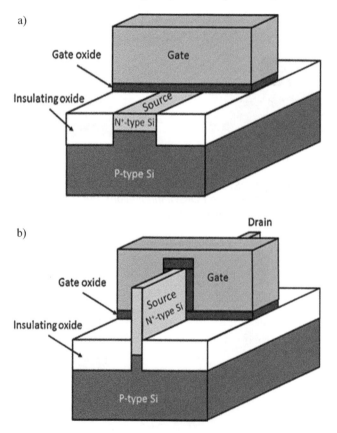

Figure 6.18. *(a) MOSFET transistor, planar structure. (b) FinFET FinFET-MOSFET or Tri-Gate (Johnson 2014). For a color version of this figure, see www.iste.co.uk/vignes/silicon1.zip*

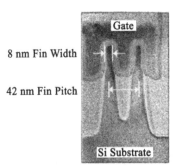

Figure 6.19. *14 nm Tri-Gate FinFET (24 nm gate length) (Intel-14 nm-IDF 14-2014)*

6.3. Appendix

	Technology node/ generation	Grid length	Channel length L_{eff}	Dielectric thickness
1960: Si-MOSFET transistor (BTL)	25 µm	25 µm		SiO_2: 100 nm
1967–1969: silicon gate technology	25 µm	25 µm		SiO_2 (60 nm)/ SiN_{34} (40 nm)
1971 → 1975	10 → 5 µm	10 → 5 µm		70 nm
1980: grid electrode: refractory metal silicide ($TiSi_2$)				
1985	2.5 µm	2.5 µm		50 nm/20 nm
1990: gate dielectric: $SiON_{xy}$ oxynitride	1 µm			
1996: CMOS component	0.25 µm	0.20 µm		5 nm
2002	130 nm	65 nm		2 nm
2002-2003: strained silicon (Si-17% Ge)	90 nm	45 nm	25 nm	1.2 nm
2007: HKMG grid/strained silicon (30% Ge)	45 nm	35 nm	25 nm	1 nm EOT HK (hafnium oxide)
2009: HKMG	32 nm	30 nm	24 nm	0.9 nm EOT
2011: FinFET	22 nm		23 nm	
2015: FD-SOI	28 nm	28 nm	24 nm	
2015: FinFET	14 nm	24 nm	22 nm	

Table 6.1. *Summary of MOSFET transistor developments*

COMMENT ON TABLE 6.1 – *Until 1997: generation technology node = length of grid electrode. After 2001: technology node ≠ gate electrode length. Table compiled from various sources: Victor Moroz, Transition from planar MOSFETs to FinFETs; Berkeley Seminar, October 28, 2011, synopsys; Chih-Tang Sah (Sah 1988).*

6.4. References

Atalla, M.M. (1965). Semiconductor devices having dielectric coatings. Patent, US3206670.

Atalla, M.M., Tannenbaum, E., Scheibner, E.J. (1959). Stabilization of silicon surfaces by thermally grown oxides. *Bell System Technical Journal*, 38(3), 749–783.

Balk, P. (1965). Effect of hydrogen annealing on silicon surfaces. In *Electrochemical Society Spring Meeting*. The Electrochemical Society, San Franscisco.

Balk, P., Burkhardt, P.G., Gregor, I.V. (1965). Orientation dependance of built-in surface charges on thermally oxidized silicon. *IEEE Proceedings*, 53(12), 2133–2134.

Bardeen, J. (1956). Semiconductor research leading to the point contact transistor. In *Great Solid State Physicists of the 20th Century*, Gonzalo, J.A. and Lopez, C.A. (eds). World Scientific, Singapore.

Berry, G. (2017). *L'Hyperpuissance de l'informatique*. Odile Jacob, Paris.

Bohr, M.T., Chau, R.S., Ghani, T., Mistry, K. (2007). The high-K solution. *IEEE Spectrum*, 44(10), 29–35.

Bois, D. and Rosencher, E. (1988). Les frontières physiques de la microélectronique. *La Recherche*, 203(10), 1176–1188.

Brown, W.L. (1953). n-Type surface conductivity on p-type germanium. *Physical Review*, 91, 518–527.

CHM (1960). The silicon engine timeline. 1960: MOS transistor demonstrated. Computer History Museum [Online]. Available at: https://www.computerhistory.org/siliconengine/metal-oxide-semiconductor-mos-transistor-demonstrated/.

CHM (1968). The silicon engine timeline. 1968: Silicon gate technology developed for ICs. Computer History Museum [Online]. Available at: https://www.computerhistory.org/siliconengine/silicon-gate-technology-developed-for-ics/.

Faggin, F., Klein, T., Vadaz, L. (1969). Insulated gate field effect transistor integrated circuits with silicon gates. *IEEE Transactions on Electron devices*, 16(2), 236.

Fenouillet-Beranger, C. (2014). Advanced planar FDSOI devices. Working document, STI/LETI 2014, 16.

George, S.M. (2010). Atomic layer deposition: An overview. *Chemical Reviews*, 110(1), 111–131.

Gusev, E.P., Cabral, C., Copel, M., Gusev, E.P., Cabral, C., Copel, M., D'Emic, C., Gribelyuk, M. (2003). Ultrathin HfO_2 films grown on silicon by atomic layer deposition for advanced gate dielectrics applications. *Microelectronic Engineering*, 69(2–4), 145–154.

Hofstein, S.R. and Heiman, F.P. (1963). The silicon insulated gate field effect transistor. *Proceedings of the IEEE*, 51(9), 1190–1202.

Hu, C.C. (2009). *Modern Semiconductor Devices for Integrated Circuits*. Prentice Hall, Upper Saddle River, NJ [Online]. Available at: https://www.chu.berkeley.edu.

Intel (2006). Intel's high-k/metal gate announcement. Intel Corporation circuits [Online]. Available at: http://www.intel.com/silicon/micron.htm.

Istratov, A.A., Heislmair, H., Weber, E.R. (2000). Iron contamination in silicon technology. *Applied Physics*, 70, 489–534.

Johnson, R.W. (2014). A brief review of atomic layer deposition: From fundamentals to applications. *Materials Today*, 17(6), 236.

Johnson, R.W., Hultquist, A., Bent, S.F. (2014). A brief review of atomic layer deposition, from fundamentals to applications. *Materials Today*, 17(5), 236–246.

Kahng, D. (1963). Electric field controlled semiconductor device. Patent, US3102230.

Kahng, D. (1976). Historical perspective on the development of MOS transistors and related devices. *IEEE Transactions on Electron Devices*, 23(7), 655–657.

Kahng, D. and Atalla, M.M. (1960). Silicon-silicon dioxide field induced surface devices. In *Solid State Research Conference*, Pittsburgh.

Kang, S. and Leblebici, Y. (2003). *CMOS Digital Integrated Circuits*, 3rd edition. McGraw Hill, New York.

Kerr, D.R. and Logan, J.S. (1964). Stabilization of SiO_2 passivation layers with P_2O_5. *IBM Journal of Research and Development*, 8(4), 376–384.

Kerwin, R.E., Klein, D.L., Sarace, J.C. (1969). Method for making MIS structure. Patent, US3475234.

Kuphaldt, T.R. (2009). *Lessons in Electric Circuits. Volume III: Semiconductors*, 5th edition. Openbook, Tutbury.

Liu, T.K. (2012). FinFET history, fundamentals and future. 2012 Symposium on VLSI technology. *Short Course*, 1–54 [Online]. Available at: https://people.eecs.berkeley.edu.

Lojek, B. (2007). *History of Semiconductor Engineering*. Springer, Berlin.

Meindl, J.D. (1977). Microelectronic circuits elements. *Scientific America*, 237(3), 13–23.

Mistry (2007). INTEL – 45nm Logic Technology [Online]. Available at: www.intel.com/technology.

Moore, G.E. (1998). The role of Fairchild in silicon technology in the early days of "Silicon Valley". *Proceedings of the IEEE*, 86(1), 53–62.

Moore, G.E., Sah, C.T., Wanlass, F.M. (1964). Metal-oxide-semiconductor field-effect devices for micropower logic circuitry. *Micropower Electronics*, 41–55.

Oldham, W.G. (1977). The fabrication of microelectronic circuits. *Scientific American*, 237(3), 41–52.

Pramanick, D. and Saxena, A.N. (1983). VLSI metallization using aluminum and its alloys. *Solid State Technology*, 26(3), 131–137.

Redoutey, J. (n.d.). Les transistors à effet de champ MOS. 5C COSP Transistor MOS. Centrale Marseille [Online]. Available at: http://jredoutey.free.fr/Puissance/3-Transistor_MOS.pdf.

Ross, I. (1957). Semiconductive translating device. Patent, US2791760.

Sah, C.T. (1961). A new semiconductor triode, the surface-potential controlled transistor. *Proceedings IRE*, 49(11), 1623–1634.

Sah, C.T. (1988). Evolution of the MOS transistor. *Proceedings of the IEEE*, 76(10), 1280–1326.

Sarace, J.C., Kerwin, R.E., Klein, D.L., Sarace, J.C., Kerwin, R.E., Klein, D.L. Edwards, R. (1968). Metal-nitride-oxide-silicon field-effect transistors, with self-aligned gates. *Solid-State Electronics*, 11(7), 653–660.

Sequeda, F.O. (1985). Integrated circuit fabrication. A process overview. *Journal of Metals*, 37, 43–50.

SIA (1999). The international technology roadmap for semiconductor [Online]. Available at: www.sematech.org/public/resources/index.htm.

Snow, E.H., Grove, A.S., Deal, B.E., Sah, C.T. (1965). Ion transport phenomena in insulating films. *Journal of Applied Physics*, 36(5), 1664–1673.

Stevenson, R. (2013). Changing the transistor channel. *IEEE Spectrum*, 6, 1–7.

Sze, S.M. (2002). *Semiconductor Devices*, 2nd edition. Wiley, New York.

Vadasz, L.L., Grove, A.S., Rowe, T.A., Moore, G.E. (1969). Silicon gate technology. *IEEE Spectrum*, 6(10), 28–35.

Wanlass, F.M. (1967). Low stand-by power complementary field effect circuitry. Patent, US3356858.

Wanlass, F.M. and Sah, C.T. (1963). Nanowatt logic using field effect metal-oxide semi-conductor triodes. In *International Solid-State Circuit Conference*. Institute of Electrical and Electronics Engineers, New York.

Wilk, G.D., Wallace, R.M., Anthony, R.M. (2001). High-gate dielectrics: Current status and materials properties considerations. *Journal of Applied Physics*, 89(10), 5243–5275.

Index of Names

Index of Terms

Summary of Volume 2

Index of Names

Index of Terms

Other titles from

in

Materials Science

2022

SALENÇON Jean — *wrong*

GADAUD Pascal
Crystal Elasticity

2019

SALENÇON Jean
Viscoelastic Modeling for Structural Analysis

2018

ABABOU Rachid
Capillary Flows in Heterogeneous and Random Porous Media 1

GODIN Nathalie, REYNAUD Pascal, FANTOZZI Gilbert
Acoustic Emission and Durability of Composite Materials
(Durability and Ageing of Organic Composite Materials Set – Volume 3)

JANNOT Yves, DEGIOVANNI Alain
Thermal Properties Measurement of Materials

2017

CAILLETAUD Georges, SAÏ Kacem, TALEB Lakhdar
Multi-mechanism Modeling of Inelastic Material Behavior

2013

BATHIAS Claude
Fatigue Limit in Metals

LEXCELLENT Christian
Shape-memory Alloys Handbook

2012

BONELLI Stéphane
Erosion of Geomaterials

BRACCINI Muriel, DUPEUX Michel
Mechanics of Solid Interfaces

DAVIM J. Paulo
Wear of Advanced Materials

DELHAES Pierre
Carbon Science and Technology: From Energy to Materials

GOUDEAU Philippe, GUINEBRETIERE René
X-Rays and Materials

OLLIVIER Jean-Pierre, TORRENTI Jean-Michel, Carcassès Myriam
Physical Properties of Concrete and Concrete Constituents

RECHO Naman
Fracture Mechanics and Crack Growth

SAANOUNI Khemais
Damage Mechanics in Metal Forming: Advanced Modeling and Numerical Simulation

VANBÉSIEN Olivier
Artificial Materials

VERDU Jacques
Oxidative Ageing of Polymers

2011

BATHIAS Claude, PINEAU André
Fatigue of Materials and Structures: Application to Damage and Design
Fatigue of Materials and Structures: Application to Design and Damage

HICHER Pierre-Yves
Multiscale Geomechanics: From Soil to Engineering Projects

IONESCU Ioan R., BOUVIER Salima, CAZACU Oana, FRANCIOSI Patrick
Plasticity of Crystalline Materials: From Dislocations to Continuum

PIJAUDIER-CABOT Gilles, DUFOUR Frédéric
Damage Mechanics of Cementitious Materials and Structures

PRIESTER Louisette
Grain Boundaries and Crystalline Plasticity

VIGNES Alain
Extractive Metallurgy 1: Basic Thermodynamics and Kinetics
Extractive Metallurgy 2: Metallurgical Reaction Processes
Extractive Metallurgy 3: Processing Operations and Routes

2010

BATHIAS Claude, PINEAU André
Fatigue of Materials and Structures: Fundamentals

CHATEIGNER Daniel
Combined Analysis

CHEVALIER Yvon, VINH TUONG Jean
Mechanical Characterization of Materials and Wave Dispersion
Mechanical Characterization of Materials and Wave Dispersion:
Instrumentation and Experiment Interpretation

DELHAES Pierre
Carbon-based Solids and Materials

2009

ALVAREZ-ARMAS Iris, DEGALLAIX-MOREUIL Suzanne
Duplex Stainless Steels

DAVIM J. Paulo
Machining Composite Materials

GALERIE Alain
Vapor Surface Treatments

Printed and bound by CPI Group (UK) Ltd, Croydon, CR0 4YY

27/10/2024

14580734-0004